TO MY FRIENDS

CONTENTS

3 The General Solutions of the Linearized Equations for Supersonic Flow

4 Boundary Conditions, Aerodynamic Forces, Uniqueness and Flow-reversal Theorems

PART II. SPECIAL METHODS

5 Subsonic Flow past Thin Bodies

6 Supersonic Flow past Nearly Plane Wings

7 Conical Fields in Supersonic Flow

8 Application of Operational Methods to Supersonic Flow

PART III. SLENDER-BODY THEORY

9 Flow at and near the Surfaces of Slender Bodies

PREFACE

The difficulty of solving the non-linear equations of motion of a compressible fluid has led to the extensive use of linear approximations to these equations in applications to aeronautics. The solutions of these linearized equations are subject to somewhat severe limitations, and a knowledge of the nature of the physical and mathematical approximations made in their derivation is helpful in assessing their applicability to particular problems. In the first part of this monograph, I have tried to set down a logical development of the theory for steady flows, giving due attention to the assumptions on which the theory is based. Perhaps an ideal course would have been to give an account of such more accurate theory as exists, and then to compare the linear approximations with it at every stage, but this would have greatly increased the size of the volume. Instead, I have quoted briefly the results of more exact theory wherever it has seemed necessary, and assumed the rest to be known or obtainable from other more complete works on the subject. Thus a knowledge of at least the elements of the theory and practice of compressible fluid flow is required from the reader. The second part deals with applications to special problems, and is mostly concerned with supersonic flows. The third part contains an account of slender-body theory, which in some ways is distinct from ordinary linearized theory and requires a separate derivation, since it is generally more accurate on and near the body and less accurate at great distances.

In developing the general theory, I have made extensive use of vectorial notation, which not only leads to a concise exposition, but, even more importantly in my opinion, helps to reveal the physical meaning of the equations. I am well aware that this will be a slight hindrance to some readers, but the notation is becoming more and more widely used, and I feel sure that the comparatively small amount of time required to learn it will be amply repaid in many other contexts. In this notation the linearized equations of motion are first-order differential equations, each of which has a simple physical interpretation, and which are shown to be integrable directly without introducing auxiliary potential functions even if they exist. The possibility thus opened up, namely to work entirely in terms of the particle velocity, appeals to me as giving a more physically satisfying approach to the subject than to work in terms of potentials. It turns out that this approach has many advantages over the more conventional treatment, and actually simplifies the analysis in many cases. Nevertheless, in order not to depart too far from standard treatments, I have exploited this possibility only in some parts of the development of the general theory, and in the chapter on conical fields. Similar considerations apply to the introduction of sources and sinks, which can have no physical reality and moreover in supersonic flow have a non-existent (infinite) mass flux unless some artifice such as that of taking the finite part of a divergent integral is employed. But the concept of sources and sinks can be a useful aid to thought, and there seems to be no point in omitting mention of these and other singularities.

As readers will rapidly become aware, the monograph shows a strong personal bias in the selection of material. This is mainly due to the fact that the monograph is a slightly revised and shortened version of an essay submitted in competition for the Adams Prize for 1949–50 in the University of Cambridge. My original intention was to expand the essay into a full-length treatise, but on consideration of the rapid rate at which the subject is developing, I have come to the conclusion that a greatly extended permanent account is not desirable at the present time. This has meant that some interesting approaches to specific problems are only mentioned briefly or are not mentioned at all, and I can only hope that this monograph with all its limitations will stimulate readers to refer to the original papers listed in the Bibliography.

I am very pleased to be able to take this opportunity of acknowledging my indebtedness to Professor S. Goldstein, Professor

M. J. Lighthill and my many other friends in the Department of Mathematics at Manchester University for the many very happy years that I spent working with them, and for all the assistance that they gave me so generously, without which it would not have been possible for me to write this monograph. My thanks are also due to Mrs D. Fahy for her careful typing of my manuscript, and to my wife for her assistance, particularly during proof-reading.

G. N. W.

CRANFIELD
August 1954

PART I. GENERAL THEORY

CHAPTER I

LINEARIZATION OF THE EQUATIONS OF MOTION

1.1. Introduction

The full treatment of the motion of a real gas consisting of discrete molecules, which themselves are complicated structures whose internal motions can affect the flow, is a problem in statistical mechanics that has not yet been solved. However, sufficient is known about the solution to make possible a critical discussion of the first simplifying assumption, that the gas can be replaced for theoretical purposes by a continuous fluid. On this assumption, the continuous fluid has at each point certain bulk or average properties of the discrete molecules which are physically measurable. These are the particle velocity, the density, the stress tensor, the coefficients of viscosity and heat conduction, the temperature, etc., to name only the more important. The pressure is derived from the stress tensor by taking the average normal stress on three mutually perpendicular surfaces at the point under consideration. The definitions of all these quantities require careful formulation, but their meanings are so well established that formal definitions are omitted here. On the assumption of statistical equilibrium, it is found that the equations of motion in terms of these average properties give an adequate representation of gaseous motions provided that the dimensions of bodies immersed in the gas, including test instruments, are large compared with the mean free path of the molecules in their vicinity, and that the times taken to complete significant changes in the state of motion of the gas are large compared with the mean time between successive collisions of one molecule with its neighbours. Thus these equations do not describe the motion of a very rarefied gas, nor do they describe accurately the motion in regions where the space or time derivatives of the physical variables are so large that the second condition above is violated. The errors arising from violations of this latter condition are not very great in general, and are usually ignored.

The effects of viscosity and heat conduction are usually of the same order of magnitude in high-speed flow, and are of significance only in regions where the space or time rates of change of velocity or temperature are large in a certain sense. Such large rates of change occur in the boundary layer at the surfaces of solid bodies, in wakes downstream from such bodies, and in the regions of nearly discontinuous velocity, etc., known as shocks or shock-waves, for example. Outside such regions the local effects of viscosity and heat conduction are small compared with the effects of inertia, and the terms involving these quantities may be neglected in the equations of motion. It must always be remembered, however, that the effects of viscosity and heat conduction in regions where they are locally important may have an important influence on the flow in regions where they are locally unimportant, as, for example, in the event of separation of the boundary layer from the surface of a body. The next simplifying assumption then is that viscosity and heat conduction can be neglected *in the equations of motion*; their effects do not disappear from the problem of determining the flow, and have to be taken into account ultimately in the form of suitably devised boundary conditions.† Thus the difficulties have only been transferred to another part of the problem of determining a flow pattern; the mathematical difficulties have been reduced at the expense of increased physical difficulties. A consequence of the neglect of viscosity is that there is no internal dissipation of energy in the fluid; also there can be no heat transfer between particles of the fluid if heat conduction is neglected, so there is thermodynamic equilibrium wherever the simplified equations of motion are valid.

The resulting equations of motion, which can be derived directly from the laws of conservation of mass, momentum and energy, are non-linear differential equations, and the mathematical difficulties associated with their solution are still formidable. However, a sufficient number of special exact solutions is known to enable a general appreciation of the behaviour of gases in high-speed flow to be obtained. Although the scope of these solutions is limited, their quantitative predictions are well confirmed by experimental measurements, thus establishing some confidence in the assumptions made above. Also, as a result of the experience gained in the

† The form of these boundary conditions is discussed in Chapter 4.

study of these solutions, qualitative pictures of gas flow under many different circumstances can be constructed, which are of great value in assessing the reasonableness or otherwise of further simplifying assumptions.

For quantitative practical applications (in the design of high-speed aircraft for example) the range of flows for which exact solutions can be obtained without undue labour is much too small. Fortunately, in many of these practical applications the resulting flows can be seen qualitatively to be such that the departures from a completely uniform flow are small, and in such cases it becomes possible to use approximate solutions. These approximate solutions can be obtained in principle either by approximating to exact solutions of the equations of inviscid flow, or by solving approximations to these equations obtained by neglecting those terms which are smaller than the others by a specified factor. These two methods are not always equivalent, even for the same problem. A first approximation is obtained by rejecting all the non-linear terms in the equations of motion and the boundary conditions, and it is the solution of the resulting linearized equations with which this monograph is concerned. It will be seen later that the linearized equations of motion are of a type which occurs frequently in mathematical physics, but some of the boundary-value problems encountered are of a kind not usually found in other branches of mathematical physics, a fact that lends considerable interest to the mathematical methods used in obtaining solutions.

Except in the beginning of the present chapter, it is assumed that the motion is steady everywhere, being flow past fixed boundaries. This assumption is not necessary, of course, but merely limits the range of motions treated. With this assumption it is still possible to treat the quasi-steady flows for which the time rates of change of all quantities are negligible in the equations of motion. It is also assumed that the fluid is not subject to any external force field: in particular, gravitational forces are neglected; their effects are very small in general, and can be inserted as small corrections if required.

1.2. Equations of motion for an inviscid non-conducting fluid

As a consequence of the assumption that a gas can be replaced by a continuous fluid, it is possible to specify a particle velocity vector, $\mathbf{u}$, a density, ρ, and a pressure, p, at every point of the fluid at which the flow is continuous; the velocity $\mathbf{u}$ is considered to be measured relative to an inertial or unaccelerated frame of reference. The state of the fluid in motion is determined completely by these three quantities, but it is convenient to introduce in addition the following thermodynamic variables: the specific internal energy, e, the specific enthalpy, h, the specific entropy, η, and the temperature, T. Except where a definite statement is made to the contrary, it is assumed in what follows that the above quantities are single-valued, bounded and continuous functions of the space co-ordinates and the time, t, and have bounded and continuous derivatives of a sufficiently high order to ensure the validity of all the formulae in which they occur.

It is shown in text-books on fluid mechanics that the kinematical equation expressing conservation of mass (the equation of continuity) is

$$\frac{\partial \rho}{\partial t}+\nabla.(\rho\mathbf{u})=0, \tag{1.2.1}$$

and that the dynamical equation expressing d'Alembert's principle for an inviscid fluid under no external forces (Euler's equation of motion) is

$$\frac{\partial \mathbf{u}}{\partial t}+\mathbf{u}.\nabla\mathbf{u}+\frac{1}{\rho}\nabla p=0. \tag{1.2.2}$$

If $\mathrm{D}/\mathrm{D}t$ denotes the operation of differentiating with respect to the time following a particle of the fluid, so that

$$\frac{\mathrm{D}}{\mathrm{D}t}\equiv\frac{\partial}{\partial t}+\mathbf{u}.\nabla, \tag{1.2.3}$$

then these equations can be written in the alternative forms

$$\frac{\mathrm{D}\rho}{\mathrm{D}t}+\rho\nabla.\mathbf{u}=0, \tag{1.2.4}$$

and

$$\frac{\mathrm{D}\mathbf{u}}{\mathrm{D}t}+\frac{1}{\rho}\nabla p=0, \tag{1.2.5}$$

respectively.

Multiplication of (1.2.1) by $\mathbf{u}$, and addition of (1.2.2) multiplied by ρ, yields the equation

$$\frac{\partial}{\partial t}(\rho\mathbf{u})+\nabla.(\rho\mathbf{u}\mathbf{u}+p\mathbf{I})=0, \tag{1.2.6}$$

where $\mathbf{I}$ is the idemtensor; if this equation is now integrated over a volume of fluid, V, bounded by a closed surface, S, that is fixed relative to the reference frame, and the integral of the second term is transformed by the divergence theorem, then it gives

$$\int_V \frac{\partial}{\partial t}(\rho \mathbf{u})\,\mathrm{d}V + \int_S (\rho \mathbf{u}\mathbf{u}.\mathbf{n} + p\mathbf{n})\,\mathrm{d}S = 0, \qquad (1.2.7)$$

where $\mathbf{n}$ is the unit outward normal to S, a result known as Euler's momentum theorem.

For mathematical purposes it is sometimes convenient to assume that fluid can be created (by sources) or destroyed (by sinks) at points in the fluid; such sources and sinks can be discrete or continuously distributed. Clearly, sources and sinks can have no physical reality, so they can occur only outside the physical boundaries of the fluid; for example, solid bodies are often replaced by an equivalent volume of fluid, and sources and sinks can arise as mathematical devices inside such bodies. The distributions postulated in practice are usually line or surface distributions, but these are most simply dealt with by taking limiting cases of volume distributions. If fluid is being created at a mass rate $\rho_0 Q$ per unit volume, then $\rho_0 Q$ is called the source density.† When sources are present, the equation of continuity (1.2.1) has to be modified by adding the term $\rho_0 Q$ to the right-hand side; the modified equation of continuity is then

$$\frac{\partial \rho}{\partial t} + \nabla.(\rho \mathbf{u}) = \rho_0 Q. \qquad (1.2.8)$$

In the general discussion of the linearized equations later in this chapter, it is assumed implicitly that Euler's equation (1.2.2) remains valid inside a source distribution. This assumption has no physical significance, since such a flow cannot be physically real, and since the source distributions encountered in practical problems are line and surface distributions at which the differential equations of motion are not valid in any case.

The fundamental thermodynamic assumptions for fluid flows are that there is local thermodynamic equilibrium at all points in the fluid, and that the same equations of state relate the local thermodynamic variables at all points. For a homogeneous fluid consisting of only one chemical substance, or of constant proportions of different chemical substances, the thermodynamic state

† The factor ρ_0, which has the dimensions of a mass density, is introduced for later convenience.

of the fluid is determined completely when any two of the thermodynamic variables are known, all others then being determinate. In fluid mechanics, it is usual to take the density, ρ, and the specific entropy, η, as the primary variables; other quantities can then be expressed as functions of ρ and η only. In particular, the pressure, p, is a function of ρ and η, and

$$p=f(\rho,\eta) \tag{1.2.9}$$

can be used as the basic equation of state; for all real fluids

$$\left(\frac{\mathrm{d}p}{\mathrm{d}\rho}\right)_\eta=\frac{\partial f}{\partial \rho}>0. \tag{1.2.10}$$

Of a small quantity of heat, δQ, supplied to unit mass of the fluid, part goes to increase the internal energy and the remainder does work in expanding the fluid; thus

$$\delta Q=\mathrm{d}e+p\,\mathrm{d}(1/\rho).\dagger \tag{1.2.11}$$

If the heat is supplied reversibly, then

$$\delta Q=T\,\mathrm{d}\eta, \tag{1.2.12}$$

and it follows from (1.2.11) that

$$\mathrm{d}e=T\,\mathrm{d}\eta-p\,\mathrm{d}(1/\rho), \tag{1.2.13}$$

and hence that p and T are related to e by the equations

$$p=\rho^2\left(\frac{\mathrm{d}e}{\mathrm{d}\rho}\right)_\eta, \quad T=\left(\frac{\mathrm{d}e}{\mathrm{d}\eta}\right)_\rho. \tag{1.2.14}$$

Under adiabatic conditions (when no heat is supplied from external sources), since there is no internal generation of heat by viscosity and no heat transfer by conduction, $\delta Q=0$ at all points of the fluid, and hence from (1.2.12) $\mathrm{d}\eta=0$ for any given particle of fluid throughout its motion. This condition can be expressed in the alternative form

$$\frac{\mathrm{D}\eta}{\mathrm{D}t}=\frac{\partial\eta}{\partial t}+\mathbf{u}.\nabla\eta=0. \tag{1.2.15}$$

The flow is said to be isentropic.

The four equations (1.2.1), (1.2.2), (1.2.9) and (1.2.15) for the four unknown functions $\mathbf{u}$, ρ, p and η are sufficient to determine the fluid motion for suitable given initial and boundary conditions,

† All thermodynamic quantities are assumed to be measured in mechanical units.

but a general discussion of fluid motions is made easier by introducing the specific enthalpy, h, defined by

$$h = e + p/\rho. \tag{1.2.16}$$

On putting this equation into differential form and using (1.2.13) to eliminate e, it follows that

$$\mathrm{d}h = T\,\mathrm{d}\eta + (1/\rho)\,\mathrm{d}p. \tag{1.2.17}$$

All particles are assumed to have the same equations of state; hence (1.2.17) is true between any pair of particles in adjacent thermodynamic states, and in particular it is true for neighbouring particles in a continuous flow. Thus (1.2.17) can be expressed in the less general but more useful form

$$\nabla h = T\nabla\eta + (1/\rho)\nabla p. \tag{1.2.18}$$

Now Euler's equation of motion can be written

$$\frac{\partial \mathbf{u}}{\partial t} + \nabla(\tfrac{1}{2}\mathbf{u}^2) - \mathbf{u}\wedge(\nabla\wedge\mathbf{u}) + \frac{1}{\rho}\nabla p = 0, \tag{1.2.19}$$

and hence, by eliminating p from these last two equations,

$$\frac{\partial \mathbf{u}}{\partial t} + \nabla(h + \tfrac{1}{2}\mathbf{u}^2) - \mathbf{u}\wedge(\nabla\wedge\mathbf{u}) = T\nabla\eta, \tag{1.2.20}$$

a result known as Crocco's theorem.

The quantity $\boldsymbol{\zeta}$ defined by

$$\boldsymbol{\zeta} = \nabla\wedge\mathbf{u} \tag{1.2.21}$$

is the vorticity; it can be shown that a small spherical particle of the fluid if suddenly solidified by internal impulses would have an angular velocity $\frac{1}{2}\boldsymbol{\zeta}$, which gives a certain physical meaning to $\boldsymbol{\zeta}$. The importance of Crocco's theorem is that it gives a relation between the vorticity and the entropy.

1.3. Homentropic flows; Kelvin's circulation theorem

An important class of flows is that for which the specific entropy is the same for all particles of the fluid. Such flows are called homentropic flows, and they arise either when the flow is homentropic at some initial instant of time, or when all the streamlines pass through some surface on which the entropy is constant; in each case (1.2.15) ensures that the flow is homentropic at all times and everywhere provided that the flow is continuous.

The circulation, K, round a closed curve, C, lying entirely in the fluid is

$$K=\int_C \mathbf{u}.\mathrm{d}\mathbf{s}, \tag{1.3.1}$$

where ds is an element of arc of C. If C is moving with the fluid, then, from (1.2.2), (1.2.18) and (1.3.1), the rate of change of K with time, denoted by $\mathrm{D}K/\mathrm{D}t$, is quite easily shown to be

$$\frac{\mathrm{D}K}{\mathrm{D}t}=\int_C T\nabla\eta.\mathrm{d}\mathbf{s}, \tag{1.3.2}$$

all other terms in the line integral vanishing when integrated round the closed curve.

For homentropic flows, $\nabla\eta=0$, and hence

$$\frac{\mathrm{D}K}{\mathrm{D}t}=0, \tag{1.3.3}$$

which is the mathematical expression of Kelvin's circulation theorem. However, $\nabla\eta=0$ is not the most general condition for which Kelvin's theorem (1.3.3) is true; for if p is a single-valued function of ρ only, in which case the flow is said to be barotropic, then $T\nabla\eta$ is the gradient of a single-valued function of ρ (or p) only, and vanishes on being integrated round the closed curve C in (1.3.2). It is shown later (§ 1.10) that for the linearized theory of nearly uniform flows, the theorem is true without any restrictions.

If S is any surface spanning C and lying entirely in the fluid, then Stokes's theorem applied to the line integral in (1.3.1) gives

$$K=\int_S \boldsymbol{\zeta}.\mathbf{n}\,\mathrm{d}S, \tag{1.3.4}$$

showing how the circulation is related to the vorticity.

1.4. Irrotational homentropic flows

In many applications, particularly to aeronautical problems, the fluid is either initially at rest, or all the particle paths originate in a uniform flow. Under such circumstances the conditions for homentropic flow are clearly satisfied, and, since the vorticity is zero in the fluid at rest or in the uniform flow, it follows from Kelvin's theorem and (1.3.4) that the vorticity is always zero for a continuous flow, which is then called irrotational. When

$$\boldsymbol{\zeta}=\nabla\wedge\mathbf{u}=0$$

the velocity can be expressed as the gradient of a scalar velocity potential, Φ, that is,

$$\mathbf{u} = \nabla\Phi. \tag{1.4.1}$$

Φ is arbitrary to the extent of a function of the time.

For irrotational homentropic flows, Crocco's equation (1.2.20) becomes

$$\nabla\left(\frac{\partial\Phi}{\partial t} + h + \tfrac{1}{2}\mathbf{u}^2\right) = 0, \tag{1.4.2}$$

which can be integrated to give Bernoulli's equation

$$\frac{\partial\Phi}{\partial t} + h + \tfrac{1}{2}\mathbf{u}^2 = \text{constant}, \tag{1.4.3}$$

where an arbitrary function of the time has been absorbed in $\partial\Phi/\partial t$. For isentropic flows, (1.2.17) gives

$$h = \int \frac{\mathrm{d}p}{\rho} = \int c^2 \frac{\mathrm{d}\rho}{\rho}, \tag{1.4.4}$$

where c is a real quantity given by

$$c^2 = \left(\frac{\mathrm{d}p}{\mathrm{d}\rho}\right)_\eta > 0 \tag{1.4.5}$$

(cf. (1.2.10)), and having the dimensions of a velocity.

To determine the equation satisfied by Φ, ρ and h can be eliminated between (1.2.1), (1.4.3) and (1.4.4), which gives

$$c^2\nabla^2\Phi - \frac{\partial^2\Phi}{\partial t^2} = \frac{\partial}{\partial t}(\nabla\Phi)^2 + \nabla\Phi.(\nabla\Phi.\nabla\nabla\Phi). \tag{1.4.6}$$

Relative to a frame of reference moving with a constant velocity equal to the velocity of a given particle of the fluid, the fluid velocity in the immediate neighbourhood of the particle differs only slightly from zero, and the equation for Φ in this neighbourhood is obtained from (1.4.6) by neglecting all the terms on the right-hand side and giving c its local value at the particle. This equation is the wave equation for disturbances propagated with speed c. Since this equation is linear in Φ, it is also the equation satisfied by the potential of a small virtual disturbance superimposed on the original flow; thus c is the speed of propagation of disturbances of very small amplitude relative to the fluid particle, that is, the local acoustic speed or local speed of sound.

1.5. Steady flows; Bernoulli's equation

In the case of steady flows, when all flow variables are independent of time, an equation for the particle velocity, $\mathbf{u}$, can be obtained by eliminating ρ and p from (1.2.1), (1.2.2) and (1.4.4) which holds on a given streamline because the flow is isentropic (though not necessarily homentropic); this gives

$$c^2\nabla.\mathbf{u}=\mathbf{u}.(\mathbf{u}.\nabla\mathbf{u})=\mathbf{u}.\nabla(\tfrac{1}{2}\mathbf{u}^2). \tag{1.5.1}$$

For steady flows, Crocco's equation (1.2.20) becomes

$$\nabla(h+\tfrac{1}{2}\mathbf{u}^2)-\mathbf{u}\wedge\boldsymbol{\zeta}=T\nabla\eta, \tag{1.5.2}$$

and after scalar multiplication by $\mathbf{u}$, this integrates immediately in isentropic flow to give

$$h+\tfrac{1}{2}\mathbf{u}^2=h_*=\text{constant on a streamline}. \tag{1.5.3}$$

The quantity h_* is the stagnation value of the specific enthalpy, or the total energy per unit mass on a streamline in steady flow, and (1.5.3), Bernoulli's equation, expresses the conservation of energy for a fluid particle.

Bernoulli's equation can be expressed in a number of useful alternative forms. By using the expression for h given in (1.4.4), (1.5.3) becomes

$$\int\frac{\mathrm{d}p}{\rho}+\tfrac{1}{2}\mathbf{u}^2=h_*, \tag{1.5.4}$$

which is the most useful form for later linearization. When the equation of state is given, the equation can take more definite forms; for a perfect gas with constant specific heats whose ratio is γ (the adiabatic index), the equation of state is

$$p=\rho^\gamma\exp(\eta/c_v), \tag{1.5.5}$$

where c_v is the specific heat at constant volume, and the specific enthalpy is

$$h=\frac{\gamma}{\gamma-1}\frac{p}{\rho}=\frac{c^2}{\gamma-1}; \tag{1.5.6}$$

Bernoulli's equation then takes the forms

$$\frac{\gamma}{\gamma-1}\frac{p}{\rho}+\tfrac{1}{2}\mathbf{u}^2=\frac{\gamma}{\gamma-1}\frac{p_*}{\rho_*} \tag{1.5.7}$$

and

$$c^2+\tfrac{1}{2}(\gamma-1)\,\mathbf{u}^2=c_*^2, \tag{1.5.8}$$

where the suffix $_*$ denotes the total or stagnation value of the quantity on the given streamline. The latter form (1.5.8) gives the relation between the local acoustic speed and the stream speed on the streamline, and in combination with (1.5.1) gives

$$c_*^2 \nabla . \mathbf{u} = \mathbf{u} . (\mathbf{u} . \nabla \mathbf{u}) + \tfrac{1}{2}(\gamma - 1) \mathbf{u}^2 \nabla . \mathbf{u} \qquad (1.5.9)$$

as an equation satisfied by the particle velocity $\mathbf{u}$ in a steady isentropic flow of a perfect fluid.

At points where the particle velocity $\mathbf{u}$ has the same magnitude as the local acoustic speed, the flow is said to be sonic. If the sonic speed is denoted by c_s, then where $|\mathbf{u}| < c_s$, the flow is called subsonic, and where $|\mathbf{u}| > c_s$, the flow is called supersonic. It follows from (1.5.8) that for subsonic flow $\mathbf{u}^2 < c_s^2 < c^2$, and for supersonic flow $\mathbf{u}^2 > c_s^2 > c^2$; hence $\mathbf{u}^2/c^2$ is less than unity for subsonic flow, and greater than unity for supersonic flow. The quantity

$$M = |\mathbf{u}|/c \qquad (1.5.10)$$

is called the local Mach number at any point, and M is less than, equal to, or greater than unity for subsonic, sonic, or supersonic flow respectively. It is to be noticed that the Mach number and the designation of the flow are defined with respect to a certain frame of reference, and that they can be changed at will simply by changing to a new frame of reference that is moving relatively to the original one.

1.6. Characteristic surfaces in steady irrotational flows

In a continuous flow, the normal derivative of the velocity can be discontinuous at certain surfaces, the tangential derivatives being continuous there. This is possible only if the equations of motion are expressible in terms of tangential derivatives, the normal derivatives being absent. Suppose that such a surface, S, exists, and let $\mathbf{n}$ be the unit normal to S at any point.

For steady irrotational flow, the equation of motion (1.4.6) becomes

$$c^2 \nabla^2 \Phi = \nabla \Phi . (\nabla \Phi . \nabla \nabla \Phi), \qquad (1.6.1)$$

which is equivalent to

$$c^2 \nabla . \mathbf{u} = \mathbf{u} . (\mathbf{u} . \nabla \mathbf{u}) \quad \text{and} \quad \nabla \wedge \mathbf{u} = 0; \qquad (1.6.2)$$

a linear combination of these equations is

$$c^2 \nabla . \mathbf{u} - \mathbf{u} . (\mathbf{u} . \nabla \mathbf{u}) + \boldsymbol{\lambda} . \nabla \wedge \mathbf{u} = 0, \qquad (1.6.3)$$

where $\boldsymbol{\lambda}$ is a vector to be determined by the condition that (1.6.3) contains only tangential derivatives of $\mathbf{u}$ on S. This last equation can be written as

$$(c^2\mathbf{I}-\mathbf{uu}+\mathbf{I}\wedge\boldsymbol{\lambda}):\nabla\mathbf{u}=0,\dagger \tag{1.6.4}$$

and the condition for normal derivatives to be absent is

$$(c^2\mathbf{I}-\mathbf{uu}+\mathbf{I}\wedge\boldsymbol{\lambda}).\mathbf{n}=0,$$

or

$$c^2\mathbf{n}-\mathbf{uu}.\mathbf{n}+\boldsymbol{\lambda}\wedge\mathbf{n}=0. \tag{1.6.5}$$

This equation for $\boldsymbol{\lambda}$ has a solution if and only if the vector $c^2\mathbf{n}-\mathbf{uu}.\mathbf{n}$ is perpendicular to $\mathbf{n}$, that is,

$$c^2=(\mathbf{u}.\mathbf{n})^2, \tag{1.6.6}$$

which is a condition for the existence of S. If the equation for S is

$$G=0, \tag{1.6.7}$$

where G is a function of position in space, then (1.6.6) gives the differential equation for G as

$$c^2(\nabla G)^2-(\mathbf{u}.\nabla G)^2=0 \quad \text{on} \quad G=0. \tag{1.6.8}$$

Surfaces for which (1.6.6) or (1.6.8) are satisfied are called characteristic surfaces of the equation of motion. These characteristic surfaces play an important part in the theory of partial differential equations, as is shown later in connexion with the linearized equations of motion. It is clear from (1.6.6) that the normal component of velocity at a characteristic surface is equal to the local speed of sound, from which it follows that real characteristic surfaces exist in steady flows only when the motion is supersonic, and that they have the important physical property of being wave fronts of steady acoustical disturbances.

Second-order partial differential equations, such as (1.6.1), or systems of first-order equations, such as (1.6.2), are classed as elliptic or hyperbolic according to whether the characteristic surfaces are imaginary or real; thus (1.6.1) is elliptic for subsonic flow and hyperbolic for supersonic flow.

1.7. Surfaces of discontinuity in steady flows

It is not always the case that the flow is continuous as has been assumed above, and on surfaces at which the velocity is discontinuous the differential equations of motion are not valid, so new equations relating the variables on different sides of such surfaces

† The symbol : denotes the double inner product of the two tensors, e.g. $\mathbf{ab}:\mathbf{cd}=\mathbf{a}.\mathbf{d}\,\mathbf{b}.\mathbf{c}$. $\mathbf{I}$ is the idemtensor as before.

are required. Let S be a surface fixed in a steady flow, let suffix 1 denote the values of variables on the upstream side of S when there is flow through S; let suffix 2 denote the values on the other (downstream) side of S, and let $\mathbf{n}$ be the unit normal to S at any point. Then conservation of mass at S gives

$$\rho_1 \mathbf{u}_1 . \mathbf{n} = \rho_2 \mathbf{u}_2 . \mathbf{n}, \tag{1.7.1}$$

and conservation of linear momentum gives

$$p_1 \mathbf{n} + \rho_1 \mathbf{u}_1 \mathbf{u}_1 . \mathbf{n} = p_2 \mathbf{n} + \rho_2 \mathbf{u}_2 \mathbf{u}_2 . \mathbf{n}. \tag{1.7.2}$$

On multiplying (1.7.2) vectorially by $\mathbf{n}$, and using (1.7.1), it gives

$$\rho_1 \mathbf{u}_1 . \mathbf{n} (\mathbf{u}_1 - \mathbf{u}_2) \wedge \mathbf{n} = 0 \tag{1.7.3}$$

as a necessary condition on the velocities for the existence of a discontinuity in a steady inviscid flow. This condition shows that there are two non-trivial possibilities, the first being

$$\mathbf{u}_1 . \mathbf{n} = \mathbf{u}_2 . \mathbf{n} = 0, \quad (\mathbf{u}_1 - \mathbf{u}_2) \wedge \mathbf{n} \neq 0. \tag{1.7.4}$$

In this case (1.7.2) gives $\quad p_1 = p_2. \qquad (1.7.5)$

Thus there is no flow through the surface and the pressure is continuous there; the surface is then a contact discontinuity or vortex sheet, being equivalent to a surface distribution of vorticity. This kind of discontinuity can occur in all types of flow, and does not invalidate the equations of motion on streamlines because no streamlines cross it.

The second possibility is that

$$(\mathbf{u}_1 - \mathbf{u}_2) \wedge \mathbf{n} = 0, \quad (\mathbf{u}_1 - \mathbf{u}_2) . \mathbf{n} \neq 0, \tag{1.7.6}$$

in which case the tangential component of velocity is continuous, and there is discontinuous flow through the surface; it follows from (1.7.1) and (1.7.2) that the density and pressure are also discontinuous. Such discontinuities are called shocks or shock waves. Conservation of energy requires the total energy per unit mass to be the same on both sides of the surface at any point, which means that Bernoulli's equation (1.5.3) holds through the surface; for a perfect gas with constant specific heats, this is

$$\frac{\gamma}{\gamma - 1} \frac{p_1}{\rho_1} + \tfrac{1}{2} \mathbf{u}_1^2 = \frac{\gamma}{\gamma - 1} \frac{p_2}{\rho_2} + \tfrac{1}{2} \mathbf{u}_2^2. \tag{1.7.7}$$

According to these equations, the discontinuity at a shock may be either compressive or expansive, but it can be shown that the entropy is discontinuous at a shock, increasing in the direction for which the velocity decreases and the pressure and density increase, and vice versa; hence, from the second law of thermodynamics, shocks can exist physically only when the discontinuity is compressive, and then the flow upstream from the shock must be supersonic. In consequence of the entropy change, if the flow is homentropic upstream from the shock, then in general it is not homentropic downstream from the shock, and in general an irrotational flow does not remain irrotational after passing through a shock.

Some further information on shocks is given in Chapter 4, but a complete discussion of their properties is outside the scope of this monograph.†

In the limit as the velocity discontinuity tends to zero, it is a simple consequence of (1.7.1) and (1.7.2) that the normal component of velocity is equal to the local speed of sound, so the shock coincides with a characteristic surface, in agreement with the definition of local speed of sound as the speed of propagation of very weak disturbances.

1.8. The genesis of linearized theory

The simplest equations for the steady flow of a perfect fluid, from a mathematical point of view, are obtained when the motions are irrotational and homentropic, and, as has been mentioned already in § 1.4, this situation arises in many problems of practical importance. In these cases, the particle velocity is the gradient of a scalar velocity potential, Φ, and (1.5.9) becomes

$$c_*^2 \nabla^2\Phi = \nabla\Phi . (\nabla\Phi . \nabla\nabla\Phi) + \tfrac{1}{2}(\gamma - 1)(\nabla\Phi)^2 \nabla^2\Phi, \qquad (1.8.1)$$

where c_* is now constant over the whole flow field under consideration. In a few cases, exact solutions of this non-linear equation (1.8.1) have been found by special methods, and there are, of course, numerical methods for solving the equation to any desired order of accuracy either by successive approximation or by step-by-step processes, and much valuable information has been obtained in this way; but in general it appears that there are only two systematic methods by means of which analytic solutions may be

† For a very full account of the properties of shocks, the reader is referred to *Supersonic Flow and Shock Waves*, by R. Courant and K. O. Friedrichs (Interscience Publishers, New York, 1948).

obtained. Both methods involve replacement of (1.8.1) by a linear equation or a system of linear equations, and no more direct analytical methods are available in general.

For flow in two dimensions it is possible to linearize (1.8.1) by the hodograph transformation, by which the velocity components replace the space co-ordinates as independent variables; the resulting equation can then be solved as a series whose terms involve hypergeometric functions. This method is outside the scope of the present monograph and is not considered in further detail here.

The second systematic method for solving is to try to expand Φ as a power series in some variable or parameter, and to determine the coefficients in this series from the boundary conditions as the solution proceeds. If the expansion is properly chosen, this process involves the solution of a sequence of inhomogeneous linear differential equations, whose inhomogeneous terms depend only upon the solutions of previous equations in the sequence.† The convergence of the series obtained in this way is usually difficult to establish. It is usually assumed that the series does converge for sufficiently small values of the variable or parameter, and that the first few terms give a useful approximation to the exact solution; but the convergence may not be uniform, a difficulty that can sometimes be overcome by modifying the expansion slightly at the region of non-uniform convergence.

The usual form of the linearized theory of steady high-speed flow treats the problem of determining the flow pattern when a fixed body is introduced into a fluid stream which was previously uniform, with constant velocity, density, and pressure everywhere. Such a body is assumed to have a surface which makes a small angle at all points with the direction of the undisturbed stream; the perturbations caused by the introduction of the body are then assumed to be so small everywhere in the field of flow that squares and higher powers of the perturbation velocity components and their derivatives can be neglected in the equations of motion and in the boundary conditions. On these assumptions a linear equation is obtained for the perturbation velocity potential, which can then be

† A slightly different process is to use successive approximations, for which the calculations are made as accurately as possible at each stage, but the principles are the same, and a solution of series type is again the result.

solved for the boundary conditions appropriate to the problem This equation is actually the first equation of a sequence as described above, the expansion chosen being in powers of a thickness parameter, t, that is related to the geometrical characteristics of the body under consideration. A common definition of t is the maximum thickness of the body perpendicular to the stream direction divided by the length of the body along the stream direction, with the proviso that the angle between the surface and the direction of the undisturbed stream is everywhere of the same order as t; but this definition may not always be applicable and others have to be used. The above form of linearized theory is the one treated in this monograph, but clearly it is not the only possible linearization (in fact, the equations for small perturbations on any given initial flow will be linear); it is distinguished from all other linearized theories by the fact that the linear differential equations have constant coefficients, which makes it the simplest mathematically.†

In order to derive the sequence of equations for the coefficients in a series expansion of Φ in powers of t, consider a uniform stream with velocity U, constant density, ρ_0, and constant pressure, p_0, at a great distance upstream from a fixed body. Then if the flow is continuous, it is irrotational and homentropic. The local speed of sound in the undisturbed stream, c_0, is given by

$$c_0^2 + \tfrac{1}{2}(\gamma - 1)\, U^2 = c_*^2, \qquad (1.8.2)$$

and the equation for Φ is then

$$c_0^2 \nabla^2\Phi = \nabla\Phi . (\nabla\Phi . \nabla\nabla\Phi) + \tfrac{1}{2}(\gamma - 1)\,[(\nabla\Phi)^2 - U^2]\,\nabla^2\Phi. \qquad (1.8.3)$$

Let x, y, z be cartesian co-ordinates for a right-handed orthogonal system of co-ordinate axes that is fixed relative to the body and is oriented so that the z-axis is parallel to the direction of the undisturbed stream,‡ and suppose that a series expansion for Φ of the form

$$\Phi = U(z + \phi_1 t + \phi_2 t^2 + \phi_3 t^3 + \ldots) \qquad (1.8.4)$$

is possible. If M_0 is the Mach number of the undisturbed stream, so that

$$M_0 = U/c_0, \qquad (1.8.5)$$

† An example of a linearized equation for perturbations on a non-uniform flow is given in Appendix 2.

‡ This system of rectangular co-ordinates is standard throughout this monograph, and is not redefined every time it is used; the unit vectors **i**, **j**, **k** parallel to the co-ordinate axes of x, y, z respectively always refer to these axes, so **k** is always the unit vector parallel to the direction of the undisturbed stream.

then, on substituting the series (1.8.4) in (1.8.3) and equating the coefficients of successive powers of t, the following sequence of equations is obtained:

$$\nabla^2\phi_1 - M_0^2\frac{\partial^2\phi_1}{\partial z^2} = 0, \tag{1.8.6}$$

$$\nabla^2\phi_2 - M_0^2\frac{\partial^2\phi_2}{\partial z^2} = M_0^2\frac{\partial}{\partial z}(\nabla\phi_1)^2 + (\gamma-1)M_0^2\frac{\partial\phi_1}{\partial z}\nabla^2\phi_1, \tag{1.8.7}$$

$$\begin{aligned}\nabla^2\phi_3 - M_0^2\frac{\partial^2\phi_3}{\partial z^2} &= M_0^2\left[2\frac{\partial}{\partial z}(\nabla\phi_1.\nabla\phi_2) + \nabla\phi_1.(\nabla\phi_1.\nabla\nabla\phi_1)\right]\\ &\quad + (\gamma-1)M_0^2\left[\frac{\partial\phi_1}{\partial z}\nabla^2\phi_2 + \frac{\partial\phi_2}{\partial z}\nabla^2\phi_1 + \tfrac{1}{2}(\nabla\phi_1)^2\nabla^2\phi_2\right],\end{aligned} \tag{1.8.8}$$

etc., from which, for given boundary conditions, the coefficients $\phi_1, \phi_2, \phi_3, \ldots$ may be determined successively. The pressure can be calculated from Bernoulli's equation as a power series in t, and is quite easily shown to be given by

$$\begin{aligned}\frac{p-p_0}{\frac{1}{2}\rho_0 U^2} &= -2\frac{\partial\phi_1}{\partial z}t - \left[2\frac{\partial\phi_2}{\partial z} + (\nabla\phi_1)^2 - M_0^2\left(\frac{\partial\phi_1}{\partial z}\right)^2\right]t^2\\ &\quad - \left[2\frac{\partial\phi_3}{\partial z} + 2\nabla\phi_1.\nabla\phi_2 - 2M_0^2\frac{\partial\phi_1}{\partial z}\frac{\partial\phi_2}{\partial z} - M_0^2\frac{\partial\phi_1}{\partial z}(\nabla\phi_1)^2\right]t^3 - \ldots.\end{aligned} \tag{1.8.9}$$

The coefficient ϕ_1, satisfying (1.8.6), gives the linearized perturbation velocity field, and (1.8.6) can also be obtained by writing

$$\mathbf{u} = U(\mathbf{k} + \nabla\phi), \quad \phi = \phi_1 t, \tag{1.8.10}$$

substituting in (1.5.1), and neglecting squares and higher powers of the derivatives of ϕ, as described above. The more elaborate analysis involving expansions in powers of t has been given to emphasize the fact that ϕ is only the first term in a series, because this is often forgotten in applications. The series must converge if its first term is to have any real meaning, and it does not converge uniformly near $M_0 = 1$ and for large M_0; it sometimes happens that plausible answers can be obtained from linearized theory for these ranges of M_0, but these are theoretically meaningless although they may have some empirical value.

The simple power series (1.8.4) is not always a possible representation of Φ. For example, when considering axially symmetrical flow past a body of revolution a series of powers of t and $\log t$ is required. The appropriate expansion in this case is of the form

$$\Phi = U[z + \phi_2 t^2 + \phi_3 t^3 + \phi_4 t^4 + \phi_5 t^5 + \phi_6 t^6 + \dots$$
$$+ \phi_4' t^4 \log t + \phi_6' t^6 \log t + \dots + \phi_6'' t^6 (\log t)^2 + \dots]. \quad (1.8.11)$$

On substituting this expansion in the equation for Φ and equating coefficients of powers of t and $\log t$, it is found that ϕ_2, ϕ_3, ϕ_4' and ϕ_6'' all satisfy the homogeneous equation (1.8.6), while the remaining coefficients satisfy inhomogeneous equations (Broderick, 1949). Other forms for the expansion containing more than one parameter, say both thickness ratio and incidence (Lighthill, 1948*a*), may be convenient, and the form that has to be used is determined ultimately by the boundary conditions on the body and the type of solution to which they lead.

1.9. Linearization of the boundary condition at a solid surface, and classification of bodies and edges

For any inviscid fluid in contact with a solid surface, it is assumed that the normal component of the relative fluid velocity at the surface is zero, and that slip may occur at the surface, so no condition is imposed on the tangential component of velocity. These conditions are discussed later, in Chapter 4, where a discussion of other boundary conditions is also given. To formulate this condition mathematically, consider any point on the surface of a body and introduce local rectangular cartesian co-ordinates ν, σ, z such that the z-axis is parallel to the direction of the undisturbed stream and the (ν, z)-plane contains the normal to the surface. In this frame of reference, let l, 0, n be the direction cosines of the outward normal from the body into the fluid at the given point. Then the boundary condition of zero normal velocity at the surface is

$$l\frac{\partial \Phi}{\partial \nu} + n\frac{\partial \Phi}{\partial z} = 0. \quad (1.9.1)$$

This boundary condition can be simplified in applications to linearized theory, and the extent to which the simplification can be carried depends upon the type of body under consideration. All bodies that come within the scope of linearized theory can be divided into two general classes: these two classes can be defined in a number of different but equivalent ways, the most direct definitions depending upon differences in geometrical configuration. A body whose local thickness is small in all directions perpendicular to the direction of the undisturbed stream is called a slender body here, and a body whose local thickness is small in one direction only is called a thin body. Thus, as the thickness (and incidence, camber, etc.) tends to zero, the limit of a slender body is a line, and the limit of a thin body is a surface, called the mean body surface. The mean body surface is clearly a developable surface whose generators are parallel to the direction of the undisturbed stream. An example of a slender body is a pointed body of revolution of small radius compared with its length, and with its axis at a small angle to the stream direction. Examples of thin bodies are a nearly plane wing, for which the mean surface is part of a plane, and a quasi-cylinder, whose mean surface is part of a cylinder (usually circular; see Chapter 8). The treatment of slender bodies requires an expansion for the potential in powers of t and $\log t$, as in (1.8.11), whereas for thin bodies an expansion in powers of t only, as in (1.8.4), usually suffices.

From the definition of a thin body, the actual body surface lies everywhere at a small distance, ν_1, from the mean surface, where $\nu_1 = O(t)$. If the origin of the co-ordinate axes for ν, σ, z is taken to be on the mean surface, and such that the ν-axis passes through the point of the body surface which is under consideration, then the boundary condition (1.9.1) is to be applied at $\nu = \nu_1$, $\sigma = z = 0$, and provided that Φ is not singular in a neighbourhood of the mean surface, it can be expanded in the form

$$l\left[\left(\frac{\partial \Phi}{\partial \nu}\right)_{\nu=0} + \nu_1\left(\frac{\partial^2 \Phi}{\partial \nu^2}\right)_{\nu=0} + \dots\right] + n\left[\left(\frac{\partial \Phi}{\partial z}\right)_{\nu=0} + \nu_1\left(\frac{\partial^2 \Phi}{\partial z\,\partial \nu}\right)_{\nu=0} + \dots\right] = 0. \tag{1.9.2}$$

On substitution of the series (1.8.4) for Φ, and since $l = 1 - \frac{1}{2}n^2 + \dots$, this equation can be separated into a sequence of equations each of which contains terms of the same order in t; these are

$$t\left(\frac{\partial \phi_1}{\partial \nu}\right)_{\nu=0} = -n, \tag{1.9.3}$$

$$t^2\left(\frac{\partial \phi_2}{\partial \nu}\right)_{\nu=0} = -nt\left(\frac{\partial \phi_1}{\partial z}\right)_{\nu=0} - \nu_1 t\left(\frac{\partial^2 \phi_1}{\partial \nu^2}\right)_{\nu=0}, \tag{1.9.4}$$

etc. The first of these is the fully linearized boundary condition, which is valid for all thin bodies; it is notably simpler than the full boundary condition since it involves only the component of fluid velocity normal to the stream direction evaluated on the mean body surface.

The above process is not valid for slender bodies, since Φ is singular in a neighbourhood of the mean line to which a slender body reduces in the limit, and an expansion of the form (1.9.2) is not possible. In this case an expansion of the form (1.8.11) has to be substituted directly in (1.9.1), which leads to a linearized boundary condition similar to (1.9.3), but the derivative on the left-hand side must now be evaluated on the actual body surface. The general theory for slender bodies is considered in Chapter 9.

The fully linearized boundary condition (1.9.3) for thin bodies is linear in ϕ_1 and in n, so if n is the sum of a number of distinct terms, the linearized perturbation potentials can be determined separately for each term, and the complete linearized solution obtained by adding these together. It is usual to separate the effects of incidence in this way. An incidence, α, is defined to be a right-handed rotation from a specified standard position of the body through an angle α about an axis perpendicular to the stream direction. With the usual system of right-handed co-ordinate axes for x, y, z, the x-axis may be taken as the axis of rotation, and if l, m, n are the direction cosines of the outward normal at points of the body surface in the standard position, then the direction cosines at incidence α are l, $m\cos\alpha - n\sin\alpha$, $n\cos\alpha + m\sin\alpha$, and the boundary condition at the surface is

$$l\frac{\partial\Phi}{\partial x}+(m\cos\alpha-n\sin\alpha)\frac{\partial\Phi}{\partial y}+(n\cos\alpha+m\sin\alpha)\frac{\partial\Phi}{\partial z}=0. \quad (1.9.5)$$

In linearized theory, α must be small and $O(t)$, together with n as before, and the fully linearized approximation to (1.9.5) is

$$t\left(l\frac{\partial\phi_1}{\partial x}+m\frac{\partial\phi_1}{\partial y}\right)_{\nu=0} = -(n+m\alpha), \quad (1.9.6)$$

or

$$t\left(\frac{\partial\phi_1}{\partial\nu}\right)_{\nu=0} = -(n+m\alpha). \quad (1.9.7)$$

This equation is a slight generalization of (1.9.3) and shows how the effect of incidence can be separated.

In general, further boundary conditions are required to determine the flow completely; these are considered in Chapter 4.

In the limiting case of zero thickness, all thin bodies have sharp edges, and it is convenient to classify these edges. The distinction between leading and trailing edges is well known, but for supersonic stream Mach numbers the edges can be divided into subclasses. If the component of the stream velocity normal to the edge is greater than the acoustic speed, c_0, then such an edge is called a supersonic edge, while if this component is less than c_0, the edge is called a subsonic edge. Thus there are supersonic leading edges, supersonic trailing edges, subsonic leading edges, and subsonic trailing edges, and the flows in the neighbourhoods of all these different types of edge are different. Of course in subsonic flows all edges are subsonic edges.

1.10. Linearization, and the equivalence of acoustical theory and linearized theory

It has been shown in § 1.8 how the linearized equation (1.8.6) for the velocity potential of disturbances on a uniform stream is obtained as the first equation of a sequence after expanding the full potential as a series. This procedure is always necessary for a rigorous approach to a linearized theory, but it is permissible to derive the linearized equations by the more convenient method of neglecting the squares, products and higher powers of small quantities, provided that the origin of the equations is kept firmly in mind; the linearized approximation is only the first term in a series, and the convergence of this series is determined largely by the neglected equations of the sequence, a fact that can hardly be emphasized too strongly.

The linearized equation (1.8.6) is independent of the ratio of specific heats, γ, which suggests that linearized theory is valid for fluids having more general equations of state than that for a perfect gas with constant specific heats. In this and the two following sections the linearized equations are re-derived on the assumption of a general equation of state in the form (1.2.9) for which (1.2.10) is satisfied. This derivation is carried out by vector methods, which leads to an elegant and physically significant form for the equations; the second-order equation for the scalar potential function is replaced by two first-order equations, which considerably simplifies discussion of linearized theory.

Consider small disturbances of a uniform stream having constant velocity, **U**, and constant density and pressure; let **v** be the perturbation velocity, so that

$$\mathbf{u}=\mathbf{U}+\mathbf{v}, \tag{1.10.1}$$

and let the constant values of all quantities in the uniform stream (except **u**) be denoted by suffix 0. If squares, products, and higher powers of **v**, $\rho-\rho_0$, $p-p_0$ and $\eta-\eta_0$, and their derivatives, are so small that they can be neglected in the equations of motion, then the equation of continuity, (1.2.1) or (1.2.4), becomes

$$\frac{\mathrm{D}}{\mathrm{D}t}\left(\frac{\rho-\rho_0}{\rho_0}\right)+\nabla.\mathbf{v}=0; \tag{1.10.2}$$

Euler's equation of motion, (1.2.2) or (1.2.5), becomes

$$\frac{\mathrm{D}\mathbf{v}}{\mathrm{D}t}+\nabla\left(\frac{p-p_0}{\rho_0}\right)=0; \tag{1.10.3}$$

and the entropy equation (1.2.15) becomes

$$\frac{\mathrm{D}}{\mathrm{D}t}(\eta-\eta_0)=0, \tag{1.10.4}$$

where in all three equations

$$\frac{\mathrm{D}}{\mathrm{D}t}\equiv\frac{\partial}{\partial t}+\mathbf{U}.\nabla. \tag{1.10.5}$$

From (1.2.9), by expanding as a Taylor series and retaining only the linear terms, the linearized equation of state is

$$p-p_0=\left(\frac{\mathrm{d}p}{\mathrm{d}\rho}\right)_{\eta_0}(\rho-\rho_0)+\left(\frac{\mathrm{d}p}{\mathrm{d}\eta}\right)_{\rho_0}(\eta-\eta_0)=c_0^2(\rho-\rho_0)+\left(\frac{\mathrm{d}p}{\mathrm{d}\eta}\right)_{\rho_0}(\eta-\eta_0), \tag{1.10.6}$$

c_0 being the speed of sound in the undisturbed stream.

The vorticity $\boldsymbol{\zeta}$ is given by

$$\boldsymbol{\zeta}=\nabla\wedge\mathbf{v}, \tag{1.10.7}$$

and by taking the curl of (1.10.3), it follows that

$$\frac{\mathrm{D}\boldsymbol{\zeta}}{\mathrm{D}t}=0, \tag{1.10.8}$$

so, in linearized theory, the vorticity of a given particle of fluid remains constant throughout the motion, provided that the flow is continuous. It follows from this that Kelvin's circulation theorem (§ 1.3),

$$\frac{DK}{Dt}=0, \tag{1.10.9}$$

is true in linearized theory whether or not the flow is barotropic, a result that also follows from the linearized form of (1.3.2):

$$\frac{DK}{Dt}=T_0\int_C \nabla\eta\,.\,d\mathbf{s}=0, \tag{1.10.10}$$

since η is a single-valued function.

By eliminating $\mathbf{v}$, ρ and η from the above equations, the pressure is found to satisfy the equation

$$c_0^2\nabla^2 p=\left(\frac{\partial}{\partial t}+\mathbf{U}.\nabla\right)^2 p. \tag{1.10.11}$$

The flow is now assumed temporarily to be irrotational; this restriction is not necessary, but it simplifies the following investigations of this section and the next. In § 1.12 this restriction is dropped for steady flows, and it is shown that the general linearized equations for steady rotational flows can be made formally identical with those for steady irrotational flows. The same can be done for unsteady rotational flows if required. For irrotational flows, $\mathbf{v}$ is the gradient of a scalar potential, ϕ say, that is,

$$\mathbf{v}=\nabla\phi, \tag{1.10.12}$$

and the above equations give the equation for ϕ as

$$c_0^2\nabla^2\phi=\left(\frac{\partial}{\partial t}+\mathbf{U}.\nabla\right)^2\phi, \tag{1.10.13}$$

while the pressure is given from (1.10.3) and (1.10.12) by the linearized Bernoulli's equation

$$p-p_0=-\rho_0\left(\frac{\partial\phi}{\partial t}+\mathbf{U}.\nabla\phi\right). \tag{1.10.14}$$

If the fluid is initially at rest in the undisturbed state ($\mathbf{U}=0$), then these last two equations are the ordinary acoustical equations for irrotational disturbances of very small magnitude ($|\mathbf{v}|\ll c_0$); alternatively, if the flow is a steady perturbation of a uniform stream, then (1.10.13) reduces to the linearized equation (1.8.6), and (1.10.14) is equivalent to the first approximation to (1.8.9). Thus the linearized equations for steady (or unsteady) disturbances of a uniform stream are equivalent in every way to the acoustical equations for a frame of reference moving with the constant velocity of the uniform stream. This fact is of great importance in the physical interpretation of the theory.

1.11. Linearization for steady irrotational flows

The linearized equations for unsteady flows given in the previous section are easily specialized for steady flows by omitting the partial derivatives with respect to time. It is convenient at this stage to include the possibility of source distributions in the flow; the modified equation of continuity has been given already in (1.2.8), and the linearized form of this equation is

$$\mathbf{U}.\nabla(\rho-\rho_0)+\rho_0\nabla.\mathbf{v}=\rho_0 Q. \tag{1.11.1}$$

The remaining linearized equations for steady flows are

$$\rho_0\mathbf{U}.\nabla\mathbf{v}+\nabla(p-p_0)=0, \tag{1.11.2}$$

$$\mathbf{U}.\nabla(\eta-\eta_0)=0, \tag{1.11.3}$$

and the linearized equation of state (1.10.6). Elimination of ρ, p and η from these equations gives

$$\nabla.\mathbf{v}-\mathbf{U}.(\mathbf{U}.\nabla\mathbf{v})/c_0^2=Q; \tag{1.11.4}$$

but, since

$$\mathbf{U}.(\mathbf{U}.\nabla\mathbf{v})=\nabla.(\mathbf{U}\mathbf{U}.\mathbf{v}), \tag{1.11.5}$$

this equation can be written

$$\nabla.(\mathbf{v}-\mathbf{U}\mathbf{U}.\mathbf{v}/c_0^2)=Q, \tag{1.11.6}$$

or, more briefly,

$$\nabla.\mathbf{w}=Q, \tag{1.11.7}$$

where the new vector $\mathbf{w}$ is given by

$$\mathbf{w}=(\mathbf{I}-\mathbf{U}\mathbf{U}/c_0^2).\mathbf{v}\equiv\mathbf{\Psi}.\mathbf{v}. \tag{1.11.8}$$

For source-free flows ($Q=0$), (1.11.7) shows that $\mathbf{w}$ is the flux of a conserved quantity in linearized theory, and it is shown below that, for irrotational flows, $\mathbf{w}$ replaces mass flux. For the standard system of rectangular co-ordinates, the tensor $\mathbf{\Psi}$ has components

$$\mathbf{\Psi}=\begin{bmatrix} 1 & 0 & 0 \\ 0 & 1 & 0 \\ 0 & 0 & 1-M^2 \end{bmatrix}, \tag{1.11.9}$$

and if $\mathbf{v}$ has rectangular components u, v, w, so that

$$\mathbf{v}=[u,v,w], \tag{1.11.10}$$

then

$$\mathbf{w}=[u,v,(1-M^2)w], \tag{1.11.11}$$

M being the Mach number of the undisturbed stream.

The linear equation of continuity (1.11.7) is valid whether or not the flow is irrotational, but is not in itself sufficient to determine $\mathbf{v}$ for flow past a given body. Thus a further equation is necessary, and this can be provided by the condition of irrotationality, which enables $\mathbf{v}$ to be expressed as the gradient of a potential function, that is

$$\mathbf{v}=\nabla\phi=\left[\frac{\partial\phi}{\partial x},\frac{\partial\phi}{\partial y},\frac{\partial\phi}{\partial z}\right], \qquad (1.11.12)$$

and from (1.11.7) and (1.11.11) it follows that the equation for ϕ in rectangular cartesian co-ordinates is

$$\frac{\partial^2\phi}{\partial x^2}+\frac{\partial^2\phi}{\partial y^2}+(1-M^2)\frac{\partial^2\phi}{\partial z^2}=Q. \qquad (1.11.13)$$

For source-free flows, this is the linear equation obtained previously in § 1·8 under more restrictive conditions, and it is justified ultimately by the more rigorous approach of the expansions in powers of the thickness ratio.

If it is assumed in addition that the flow is homentropic, with $\eta=\eta_0$, then the pressure can be obtained from Bernoulli's equation (1.5.4), which can be written as

$$\int_{p_0}^{p}\frac{dp}{\rho}+\tfrac{1}{2}\mathbf{u}^2=\tfrac{1}{2}\mathbf{U}^2. \qquad (1.11.14)$$

For sufficiently small departures from the undisturbed uniform flow, the integrand in (1.11.14) can be expanded as a Taylor series in powers of $(p-p_0)$, and the equation becomes

$$\int_{p_0}^{p}\left\{\frac{1}{\rho_0}-\frac{1}{\rho_0^2}\left(\frac{d\rho}{dp}\right)_{\eta_0}(p-p_0)+\ldots\right\}dp+\tfrac{1}{2}\mathbf{u}^2=\tfrac{1}{2}\mathbf{U}^2, \qquad (1.11.15)$$

which, on carrying out the integration, gives

$$\frac{p-p_0}{\rho_0}-\frac{1}{2c_0^2}\left(\frac{p-p_0}{\rho_0}\right)^2+\ldots=\tfrac{1}{2}\mathbf{U}^2-\tfrac{1}{2}\mathbf{u}^2=-\mathbf{U}.\mathbf{v}-\tfrac{1}{2}\mathbf{v}^2. \qquad (1.11.16)$$

On solving this equation for $(p-p_0)/\rho_0$ by inverting the series, an approximate equation for the pressure is determined in the form

$$\frac{p-p_0}{\rho_0}=-\mathbf{U}.\mathbf{v}-\tfrac{1}{2}\mathbf{v}.\mathbf{w}. \qquad (1.11.17)$$

This last result is called the quadratic approximation to Bernoulli's equation. If $\mathbf{v}$ is the linearized perturbation velocity, then on comparison with (1.8.9) it is apparent that (1.11.17) is not the true second approximation to the pressure, since the term equivalent to $-2(\partial\phi_2/\partial z)t^2$ in (1.8.9) is missing from (1.11.17), but it is nevertheless the expression that has to be used for the calculation of aerodynamic forces in certain problems. The linearized Bernoulli equation is

$$p-p_0=-\rho_0\mathbf{U}.\mathbf{v}, \tag{1.11.18}$$

which is completely equivalent to the first term in (1.8.9).

The density changes can be found by expanding the density, ρ, as a Taylor series in powers of $p-p_0$ as follows:

$$\rho=\rho_0+\left(\frac{d\rho}{dp}\right)_{\eta_0}(p-p_0)+\ldots=\rho_0+(p-p_0)/c_0^2+\ldots, \tag{1.11.19}$$

or, by using (1.11.18),

$$\rho-\rho_0=-\rho_0\mathbf{U}.\mathbf{v}/c_0^2 \tag{1.11.20}$$

to a linear approximation. The perturbation mass flux is

$$\begin{aligned}\rho\mathbf{u}-\rho_0\mathbf{U}&=\rho_0(1-\mathbf{U}.\mathbf{v}/c_0^2+\ldots)(\mathbf{U}+\mathbf{v})-\rho_0\mathbf{U}\\&=\rho_0(\mathbf{v}-\mathbf{U}\mathbf{U}.\mathbf{v}/c_0^2+\ldots)=\rho_0\mathbf{w}+\ldots;\end{aligned} \tag{1.11.21}$$

hence, to a linear approximation,

$$\mathbf{w}=(\rho\mathbf{u}-\rho_0\mathbf{U})/\rho_0, \tag{1.11.22}$$

which gives the direct relation between $\mathbf{w}$ and the perturbation mass flux when the flow is irrotational and homentropic.

The vector $\mathbf{w}$ was introduced in the present context by Robinson (1948*b*), who called it the reduced current velocity.

The equation for the characteristic surfaces of the linearized equations for irrotational flow is obtained as a special case of (1.6.6) in the form

$$c_0^2=(\mathbf{U}.\mathbf{n})^2 \quad \text{or} \quad \mathbf{n}.\mathbf{\Psi}.\mathbf{n}=0; \tag{1.11.23}$$

the equation for G (see (1.6.7) and (1.6.8)) in rectangular coordinates is then

$$\left(\frac{\partial G}{\partial x}\right)^2+\left(\frac{\partial G}{\partial y}\right)^2+(1-M^2)\left(\frac{\partial G}{\partial z}\right)^2=0 \quad \text{on} \quad G=0. \tag{1.11.24}$$

Thus, as is to be expected, the linear equations have real characteristic surfaces and are of hyperbolic type when the undisturbed stream is supersonic; and they are of elliptic type for subsonic streams.

In the absence of surface distributions of sources, if $\mathbf{n}$ is the unit normal to a surface at which the velocity is discontinuous, then apart from the existence of vortex sheets in the flow, $\mathbf{n}.\mathbf{w}$ and $\mathbf{n}\wedge\mathbf{v}$ must be continuous across any discontinuity. These conditions are the linearized equivalents of (1.7.1) and (1.7.6). It follows that $(\mathbf{n}.\mathbf{\Psi})\wedge(\mathbf{n}\wedge\mathbf{v})$ must also be continuous, and since

$$\begin{aligned}(\mathbf{n}.\mathbf{\Psi})\wedge(\mathbf{n}\wedge\mathbf{v}) &= (\mathbf{n}.\mathbf{\Psi}.\mathbf{v})\,\mathbf{n}-(\mathbf{n}.\mathbf{\Psi}.\mathbf{n})\,\mathbf{v}\\ &= (\mathbf{n}.\mathbf{w})\,\mathbf{n}-(\mathbf{n}.\mathbf{\Psi}.\mathbf{n})\,\mathbf{v},\end{aligned} \tag{1.11.25}$$

and on taking account of the first of the above conditions, the surface can exist if and only if

$$\mathbf{n}.\mathbf{\Psi}.\mathbf{n}=0, \tag{1.11.26}$$

which shows that such surfaces must be characteristic surfaces in linearized theory. The consequences of this are developed in §4.3.

1.12. Linearization for rotational flows

The linearized equations (1.11.1)–(1.11.11) hold for rotational flows, but in this case the condition of irrotationality (1.11.12) has to be modified. If the vorticity, $\boldsymbol{\zeta}$, is known, then the system of equations is completed by

$$\nabla\wedge\mathbf{v}=\boldsymbol{\zeta}, \tag{1.12.1}$$

but of course $\boldsymbol{\zeta}$ is not always known *a priori*. Nevertheless, it is found convenient later to take (1.11.7) and (1.12.1) as the fundamental linear equations, and it is shown in Chapters 2 and 3 that these equations can be integrated as they stand.

If the pressure is chosen as the dependent variable, then, for steady flows with sources, $\mathbf{v}$, ρ and η can be eliminated from (1.11.1), (1.11.2), (1.11.3) and (1.10.6) to give

$$\frac{\partial^2 p}{\partial x^2}+\frac{\partial^2 p}{\partial y^2}+(1-M^2)\frac{\partial^2 p}{\partial z^2}=-\rho_0 U\frac{\partial Q}{\partial z}; \tag{1.12.2}$$

which shows that p satisfies an equation of the same form as that for ϕ in irrotational flows (cf. (1.10.11) and (1.10.13) for unsteady flows).

A more convenient and illuminating linearized theory of rotational flows can be obtained by integrating (1.11.2) with respect to z from $z=-\infty$ to z, which gives

$$\mathbf{v}-\mathbf{v}_1=-\int_{-\infty}^{z}\nabla\left(\frac{p-p_0}{\rho_0 U}\right)dz, \tag{1.12.3}$$

where $\mathbf{v}_1$ is the value of $\mathbf{v}$ at $z=-\infty$, that is, at great distances upstream. Now if a scalar potential function, ϕ', is defined by

$$\phi' = -\int_{-\infty}^{z} \frac{p-p_0}{\rho_0 U}\,dz, \tag{1.12.4}$$

then (1.12.3) can be written as

$$\mathbf{v} = \mathbf{v}_1 + \nabla\phi' = \mathbf{v}_1 + \mathbf{v}', \tag{1.12.5}$$

say, and the perturbation pressure is given by

$$p - p_0 = -\rho_0 \mathbf{U}.\mathbf{v}', \tag{1.12.6}$$

showing that the pressure is independent of conditions at great distances upstream. These equations, (1.12.5) and (1.12.6), replace (1.11.12) and (1.11.18) when there is vorticity in the flow. Since $\mathbf{v}'$ is an irrotational vector, it follows that all the vorticity is contained in $\mathbf{v}_1$; $\mathbf{v}_1$ is clearly independent of z, and hence so is the vorticity, in agreement with (1.10.8) for the special case of steady flow.

Without any loss of generality, it can be assumed that $\mathbf{v}_1$ is a possible steady perturbation flow without sources, in which case (1.12.2) shows that ϕ' satisfies the same equation as ϕ, namely, (1.11.13). This result can be obtained in another way, by eliminating the z-component of $\mathbf{v}$ and ρ from the z-component of (1.11.2) and (1.11.6); in rectangular cartesian co-ordinates, the resulting equation is

$$\frac{\partial u}{\partial x} + \frac{\partial v}{\partial y} + (1-M^2)\frac{\partial}{\partial z}\left(\frac{p_0-p}{\rho_0 U}\right) = Q, \tag{1.12.7}$$

which again leads to an equation of the same form as (1.11.13) for ϕ'.

This equation (1.12.7) shows that a vector, ϖ say, with rectangular components given by

$$\varpi = [u, v, (1-M^2)(p_0-p)/\rho_0 U], \tag{1.12.8}$$

represents the flux of another conserved quantity. The vector ϖ is related to the vector $\mathbf{w}$ introduced previously in §1.11, but the relation is not a simple one. It is difficult to find an interpretation for ϖ, but the equations can be put into a slightly different form for which an interpretation can be given.

Instead of considering perturbations on a uniform stream, consider perturbations on a slightly non-uniform stream whose velocity is

$$\mathbf{u}_1 = \mathbf{U} + \mathbf{v}_1. \tag{1.12.9}$$

Let
$$\mathbf{u}=\mathbf{u}_1+\mathbf{v}', \qquad (1.12.10)$$
and generally let quantities pertaining to the initial slightly non-uniform stream be denoted by a suffix 1, and let the perturbation values of these quantities relative to the non-uniform conditions be denoted by a prime. From its definition, $\mathbf{v}_1$ is independent of z and is a possible steady perturbation of the uniform stream $\mathbf{U}$; hence it follows from (1.11.2) that p_1-p_0 is a constant, and without loss of generality this constant can be taken to be zero, so that from (1.12.6)
$$p-p_1=p-p_0=-\rho_0\mathbf{U}.\mathbf{v}'. \qquad (1.12.11)$$
The vector $\boldsymbol{\varpi}$ now becomes
$$\boldsymbol{\varpi}=\boldsymbol{\varpi}_1+\boldsymbol{\varpi}'=\boldsymbol{\varpi}_1+\mathbf{w}', \qquad (1.12.12)$$
where $\boldsymbol{\varpi}'$ has been replaced by $\mathbf{w}'$ for reasons that will become apparent; in rectangular components
$$\boldsymbol{\varpi}_1=[u_1(x,y),v_1(x,y),0], \qquad (1.12.13)$$
and
$$\begin{aligned}\mathbf{w}'&=[u',v',(1-M^2)(p_0-p)/\rho_0 U]\\&=[u',v',(1-M^2)w']=\boldsymbol{\Psi}.\mathbf{v}'. \qquad (1.12.14)\end{aligned}$$
Since the initial slightly non-uniform flow is source-free, $\boldsymbol{\varpi}_1$ satisfies
$$\nabla.\boldsymbol{\varpi}_1=0, \qquad (1.12.15)$$
and hence from (1.12.7), $\mathbf{w}'$ satisfies
$$\nabla.\mathbf{w}'=Q, \qquad (1.12.16)$$
showing that both $\boldsymbol{\varpi}_1$ and $\mathbf{w}'$ represent fluxes of conserved quantities.

The vector $\mathbf{w}'$ has an interpretation precisely similar to that found for $\mathbf{w}$ in irrotational flows; for, to a linear approximation,
$$\begin{aligned}\rho&=\rho_1+\left(\frac{\mathrm{d}\rho}{\mathrm{d}p}\right)_{\eta_1}(p-p_1)+\left(\frac{\mathrm{d}\rho}{\mathrm{d}\eta}\right)_{p_1}(\eta-\eta_1)\\&=\rho_1+(p-p_1)/c_1^2=\rho_1+(p-p_0)/c_0^2=\rho_1-\rho_0\mathbf{U}.\mathbf{v}'/c_0^2, \qquad (1.12.17)\end{aligned}$$
the entropy term disappearing because the flow is isentropic and $\eta=\eta_1$† everywhere; hence
$$\begin{aligned}\rho\mathbf{u}-\rho_1\mathbf{u}_1&=\rho_1\mathbf{v}'+(\rho-\rho_1)\mathbf{U}\\&=\rho_0\mathbf{v}'-\rho_0\mathbf{U}\mathbf{U}.\mathbf{v}'/c_0^2\\&=\rho_0\mathbf{w}'=\rho_1\mathbf{w}', \qquad (1.12.18)\end{aligned}$$
all to the same (linearized) order of approximation.

† η_1 is not constant over the flow field in general.

The above analysis leads to the important conclusion that *the linearized equations for steady irrotational flow, and their interpretations, are unaffected by small steady rotational non-uniformities in the main stream.* Of course, steady irrotational non-uniformities are included as a special case. It must be emphasized, however, that, when applying the equations, care must be taken to apply the boundary conditions for the complete velocity vector $\mathbf{U}+\mathbf{v}_1+\mathbf{v}'$, for although $\mathbf{v}_1$ makes no direct contribution to the pressure on the above assumptions (because its z-derivative is zero), it does affect the pressure ultimately by its effect on $\mathbf{v}'$ through the boundary conditions.

The application of these results presents no difficulty if $\mathbf{v}_1$ is known, but sometimes the only information available is a knowledge of the streamwise component (that is, the z-component) of vorticity;† however, the pressure can still be determined even with this somewhat meagre information. For, from (1.12.13) and (1.12.15), the rectangular components of $\mathbf{v}_1$ satisfy

$$\frac{\partial u_1}{\partial x}+\frac{\partial v_1}{\partial y}=0, \qquad (1.12.19)$$

so a scalar function ψ must exist such that

$$u_1=\frac{\partial \psi}{\partial y}, \quad v_1=-\frac{\partial \psi}{\partial x}; \qquad (1.12.20)$$

and if the z-component of vorticity is denoted by $Z(x,y)$, then

$$\frac{\partial v_1}{\partial x}-\frac{\partial u_1}{\partial y}=Z(x,y), \qquad (1.12.21)$$

and ψ satisfies
$$\frac{\partial^2 \psi}{\partial x^2}+\frac{\partial^2 \psi}{\partial y^2}=-Z(x,y). \qquad (1.12.22)$$

Thus u_1, v_1 are the components of a two-dimensional velocity field identical with that due to a two-dimensional distribution of vorticity in incompressible flow, the vortex tubes being parallel to the z-axis and of constant strength per unit area. For given vorticity, $Z(x,y)$, the velocity components u_1 and v_1 are independent of the Mach number. Although the z-component of $\mathbf{v}_1$ is not known, this does not matter, because the vector $\boldsymbol{\varpi}_1$ is completely determined, and $\mathbf{w}'$ can then be found, since boundary conditions at

† E.g. after passing through a strong curved shock.

the surfaces of solid bodies involve only the transverse or cross-stream components of velocity (that is, the x- and y-components) in linearized theory, and boundary conditions at free surfaces or vortex sheets involve only the pressure, all of which are contained in $\boldsymbol{\varpi}_1+\mathbf{w}'$; the z-component of $\mathbf{w}'$ then gives the pressure everywhere.

If $Z=0$, so that there is no streamwise component of vorticity, then the cross-stream components of $\mathbf{v}_1$ can be absorbed in $\mathbf{v}'$ without affecting the equations of motion or the pressure, which confirms Lighthill's (1949*a*) conclusion that *the pressure determined by irrotational flows is unaffected by cross-stream vorticity in linearized theory*. This important result explains, for example, why the effect of a detached bow shock is so small at some distance from a rounded leading edge in two-dimensional supersonic flow.

In the subsequent general theory of the linearized equations, they will usually be taken in the form

$$\nabla \wedge \mathbf{v}=\boldsymbol{\zeta}, \quad \nabla . \mathbf{w}=Q, \quad \mathbf{w}=\boldsymbol{\Psi} . \mathbf{v}, \qquad (1.12.23)$$

on the assumption that $\boldsymbol{\zeta}$ and Q are known. The equations

$$\nabla \wedge \mathbf{v}'=0, \quad \nabla . \mathbf{w}'=Q, \quad \mathbf{w}'=\boldsymbol{\Psi} . \mathbf{v}', \qquad (1.12.24)$$

are clearly a special case of (1.12.23).

CHAPTER 2

THE GENERAL SOLUTIONS OF THE LINEARIZED EQUATIONS FOR SUBSONIC FLOW

2.1. Reduction to Laplace's equation for irrotational flow

The equation for the perturbation velocity potential ϕ, (1.11.13), which is also satisfied by the potential ϕ' in rotational motions (§ 1.12), has solutions for $M<1$ which differ fundamentally from those when $M>1$. In this chapter the case of subsonic flow, $M<1$, is investigated in some detail; the case of supersonic flow, $M>1$, is considered in the next chapter.

When the flow is subsonic, the acoustic speed c_0 is greater than the stream speed, U, and it is apparent on physical grounds that any local disturbance caused in the fluid will ultimately produce effects in the whole fluid. For steady motions this fact is reflected in the form taken by (1.11.13) when $M<1$, which is then of elliptic type. The general theory of elliptic differential equations shows that the solutions determined by given, not necessarily analytic, boundary data on closed surfaces are analytic functions of the independent variables inside the closed surfaces except, perhaps, at isolated singularities. Moreover, if small changes are made in the boundary data at some points of the surface, then the solution is changed almost everywhere inside the surface. The specification of the form of the boundary data required to ensure existence and uniqueness of the solutions is an important matter which is dealt with later.

It is convenient to write the equation for ϕ in subsonic flows as

$$\frac{\partial^2\phi}{\partial x^2}+\frac{\partial^2\phi}{\partial y^2}+\beta^2\frac{\partial^2\phi}{\partial z^2}=Q, \tag{2.1.1}$$

where $$\beta=\sqrt{(1-M^2)}>0.$$

The affine transformation

$$x=x'/k, \quad y=y'/k, \quad z=\beta z'/k, \quad k\neq 0, \tag{2.1.2}$$

then transforms this equation to Laplace's equation

$$\frac{\partial^2\phi}{\partial x'^2}+\frac{\partial^2\phi}{\partial y'^2}+\frac{\partial^2\phi}{\partial z'^2}=0, \tag{2.1.3}$$

which is the equation satisfied by the velocity potential for the flow of an inviscid incompressible fluid, or liquid, of uniform density.

If
$$\phi=\phi_i=f(x',y',z') \tag{2.1.4}$$
represents any solution of Laplace's equation (i.e. any harmonic function) which gives the irrotational flow of an inviscid liquid past some body, then
$$\phi=\frac{C}{\beta}f(kx,ky,kz/\beta) \tag{2.1.5}$$
is a solution of (2.1.1), and gives a related linearized subsonic flow of a compressible fluid. The two parameters C and k appear to be arbitrary, but strictly this is not so, since a relation between them is imposed by the boundary condition on the body, as is shown below.

2.2. The relation between subsonic compressible flows and incompressible flows

If l, m, n are the direction cosines of the outward normal to the surface of some definite body in the compressible flow, then the boundary condition on the surface is (§ 1.9)
$$l\frac{\partial\phi}{\partial x}+m\frac{\partial\phi}{\partial y}=-nU. \tag{2.2.1}$$
For the corresponding incompressible flow, if l', m', n' are the direction cosines of the outward normal to the corresponding body, then the boundary condition for the perturbation potential ϕ_i is
$$l'\frac{\partial\phi_i}{\partial x'}+m'\frac{\partial\phi_i}{\partial y'}=-n'U, \tag{2.2.2}$$
which must be equivalent to (2.2.1). In this last equation it is assumed that n' is small compared with unity for the incompressible flow; that this is so can be seen from the relation
$$\frac{l'}{l}=\frac{m'}{m}=\frac{n'}{\beta n} \tag{2.2.3}$$
(which holds by virtue of the transformation (2.1.2)), and the fact that n is small.

From (2.1.2), (2.1.4) and (2.1.5) it follows that
$$\frac{\partial\phi}{\partial x}=\frac{Ck}{\beta}\frac{\partial\phi_i}{\partial x'},\quad \frac{\partial\phi}{\partial y}=\frac{Ck}{\beta}\frac{\partial\phi_i}{\partial y'},\quad \frac{\partial\phi}{\partial z}=\frac{Ck}{\beta^2}\frac{\partial\phi_i}{\partial z'}, \tag{2.2.4}$$

and from this and (2.2.3), the boundary condition (2.2.1) becomes

$$Ck\left(l'\frac{\partial\phi_i}{\partial x'}+m'\frac{\partial\phi_i}{\partial y'}\right)=-n'U, \tag{2.2.5}$$

which must be identical with (2.2.2). This is so if, and only if,

$$Ck=1, \tag{2.2.6}$$

which is a necessary relation between C and k in general.

The pressure in the compressible flow is given approximately by the quadratic approximation to Bernoulli's equation,

$$\frac{p-p_0}{\rho_0}=-U\frac{\partial\phi}{\partial z}-\frac{1}{2}\left(\frac{\partial\phi}{\partial x}\right)^2-\frac{1}{2}\left(\frac{\partial\phi}{\partial y}\right)^2-\frac{\beta^2}{2}\left(\frac{\partial\phi}{\partial z}\right)^2. \tag{2.2.7}$$

(The last term in this expression can always be shown to be negligible, but it is retained here for convenience.) The pressure, p_i, in the corresponding incompressible flow with the same undisturbed stream density and pressure is given by

$$\frac{p_i-p_0}{\rho_0}=-U\frac{\partial\phi_i}{\partial z'}-\frac{1}{2}\left(\frac{\partial\phi_i}{\partial x'}\right)^2-\frac{1}{2}\left(\frac{\partial\phi_i}{\partial y'}\right)^2-\frac{1}{2}\left(\frac{\partial\phi_i}{\partial z'}\right)^2. \tag{2.2.8}$$

By substituting from (2.2.4) in (2.2.7) and using (2.2.6), it is seen that

$$p-p_0=(p_i-p_0)/\beta^2 \tag{2.2.9}$$

to a quadratic approximation.

Thus in order to calculate the pressure and forces on a body in subsonic flow, the incompressible flow past a body whose thickness, incidence, camber, etc., are β times those of the original body must be calculated. The perturbation pressure in the compressible flow is then $1/\beta^2$ times the perturbation pressure at corresponding points in the incompressible flow.

This rule appears to have been given first by Göthert (1940), and re-derived independently by Sears (1946, 1947) and by Young and Kirkby (1947) for the special case of a body of revolution (the above extension follows immediately from their work); it must be used for slender bodies, but for thin bodies it turns out to be unnecessarily strict, and then C and k can be specified independently as shown below.

2.3. The distortion of the streamlines

The correspondence between subsonic compressible flows and incompressible flows past thin bodies has been worked out in some detail by Goldstein and Young (1943), and the following method was used by them in their investigations.

The differential equations of the streamlines in the compressible flow are, to a linear approximation,

$$\frac{dx}{(\partial\phi/\partial x)} = \frac{dy}{(\partial\phi/\partial y)} = \frac{dz}{U}. \qquad (2.3.1)$$

Hence the equations of the streamline which passes through $x = x_0$, $y = y_0$ at $z = -\infty$ upstream are

$$x - x_0 = \frac{1}{U}\int_{-\infty}^{z} \frac{\partial\phi}{\partial x}\,dz, \quad y - y_0 = \frac{1}{U}\int_{-\infty}^{z} \frac{\partial\phi}{\partial y}\,dz. \qquad (2.3.2)$$

Both sides of these equations are small for all values of z in linearized theory, so the integrands can be evaluated on $x = x_0$, $y = y_0$ instead of on the actual streamline, the error being of the second order of small quantities on the assumption that all higher derivatives of ϕ are bounded on and near to the streamline under consideration. This assumption is correct near thin bodies, but not near slender bodies.

The corresponding streamline in a related incompressible flow clearly passes through $x' = kx_0$, $y' = ky_0$ and its equations are

$$x' - kx_0 = \frac{1}{U}\int_{-\infty}^{z'} \frac{\partial\phi_i}{\partial x'}\,dz', \quad y' - ky_0 = \frac{1}{U}\int_{-\infty}^{z'} \frac{\partial\phi_i}{\partial y'}\,dz'. \qquad (2.3.3)$$

Then, by virtue of (2.2.4),

$$x - x_0 = C(x' - kx_0), \quad y - y_0 = C(y' - ky_0). \qquad (2.3.4)$$

The quantities $x - x_0$, $y - y_0$ give the *distortion* of the streamline $x = x_0$, $y = y_0$ in the undisturbed flow, due to the disturbance caused by the introduction of the body. Thus (2.3.4) shows that the distortion of the streamlines in the compressible flow is C times the distortion of the streamlines at the corresponding point in the related incompressible flow. On the other hand, the affine transformation (2.1.2) causes a *displacement* of the undisturbed streamlines from $x = x_0$, $y = y_0$ in the compressible flow to $x' = kx_0$, $y' = ky_0$ in the incompressible flow. Thus the distortion and displacement of the streamlines can be specified independently, through the parameters C and k respectively, for flow past thin bodies.

The actual choice of C and k is governed solely by considerations of convenience in any particular problem. For example, if the distortions are to be the same, then $C=1$, the choice of k still being open. If $k=1$, then (2.2.6) is satisfied and the rule (2.2.9) for the calculation of the pressure applies. If $k=\beta$, say, then the situation is slightly more complicated. The point in the incompressible flow corresponding to x, y, z in the compressible flow is $x'=\beta x$, $y'=\beta y$, $z'=z$, and it must be noted that these are strictly points in the undisturbed stream. The length of the body in the stream direction is unaltered and the lateral dimensions (e.g. the span of a nearly plane wing) in the incompressible flow are decreased to β times their values in the compressible flow, *but the thickness, where small, remains unchanged*, since the distortions of the streamlines are the same in both flows. To a linear approximation, the perturbation pressure in the compressible flow is $1/\beta$ times the perturbation pressure in the incompressible flow. Slender bodies, such as a body of revolution, remain undistorted in this transformation ($C=1$, $k=\beta$), and it was once thought that this is the appropriate transformation for such bodies. That this cannot be so follows from the results of § 2.2. The error here lies in the fact that the potential is singular in every cross-section of a slender body, and hence it is not true that the derivatives of the potential are bounded on the undisturbed streamline which becomes the body surface.

As another example, suppose that the circulations on corresponding circuits are required to be the same in both flows. The circulation, K, on a closed circuit in the compressible flow is

$$K=\int\left(\frac{\partial\phi}{\partial x}\,\mathrm{d}x+\frac{\partial\phi}{\partial y}\,\mathrm{d}y+\frac{\partial\phi}{\partial z}\,\mathrm{d}z\right), \qquad (2.3.5)$$

and the circulation, K_i, on the corresponding circuit in the incompressible flow is

$$K_i=\int\left(\frac{\partial\phi_i}{\partial x'}\,\mathrm{d}x'+\frac{\partial\phi_i}{\partial y'}\,\mathrm{d}y'+\frac{\partial\phi_i}{\partial z'}\,\mathrm{d}z'\right). \qquad (2.3.6)$$

By using (2.1.2) and (2.2.4), it follows from these last two equations that

$$K=CK_i/\beta. \qquad (2.3.7)$$

Hence, the circulations in any two corresponding circuits are the same if $C=\beta$. The choice of k is open, as before.

Some applications of the above method are given later in Chapter 5.

2.4. Boundary data: uniqueness

The divergence theorem applied to the vector $\phi\mathbf{w}$ inside a closed surface S, bounding a volume V, gives

$$\int_S \phi\mathbf{w}.\mathbf{n}\,\mathrm{d}S = \int_V (\phi\nabla.\mathbf{w} + \mathbf{w}.\nabla\phi)\,\mathrm{d}V, \tag{2.4.1}$$

where $\mathbf{n}$ is the outward normal to S.

If $\mathbf{w}$ satisfies $\nabla.\mathbf{w}=0$, so that ϕ satisfies (2.1.1), then the integrand of the volume integral is

$$\mathbf{v}.\mathbf{w} = u^2+v^2+\beta^2 w^2, \tag{2.4.2}$$

where u, v, w are the rectangular components of $\mathbf{v}$ as before, and (2.4.1) becomes

$$\int_S \phi\mathbf{w}.\mathbf{n}\,\mathrm{d}S = \int_V (u^2+v^2+\beta^2 w^2)\,\mathrm{d}V. \tag{2.4.3}$$

Now if the surface S is divided into two parts S_1, S_2, neither of which need be connected, such that $\phi=0$ on S_1 and $\mathbf{w}.\mathbf{n}=0$ on S_2, then the surface integral vanishes; hence the volume integral is identically zero, and

$$u=v=w=0 \quad \text{and} \quad \phi=\text{constant} \tag{2.4.4}$$

at all points of V. The potential ϕ can be defined to be continuous at all points of S and V, so, if S_1 exists, $\phi=0$ everywhere in V. Alternatively, if S_2 is the whole of S, and ϕ is not known at any point of S, then the constant in (2.4.4) is arbitrary, but can be taken to be zero.

If the volume V is bounded externally by a very large sphere S_3, of radius R, and internally by a surface $S=S_1+S_2$ of finite dimensions, and if $\phi=O(1/R)$ as $R\to\infty$,† then the surface integral over S_3 vanishes in the limit as $R\to\infty$, while the volume integral converges. Thus, with this condition on ϕ, the identity (2.4.3) holds for the unbounded volume outside S, and if $\phi=0$ on S_1, and $\mathbf{w}.\mathbf{n}=0$ on S_2, as before, the result (2.4.4) follows, except that $\phi=0$ everywhere in V always, because $\phi=0$ at infinity.

These results are so similar to the equivalent results for harmonic functions that this brief outline of their proofs will suffice here. They show that for any given set of points on S, it is sufficient to

† It can be shown that this condition must be satisfied if $\mathbf{v}\to 0$ as $R\to\infty$.

specify either ϕ or $\mathbf{w}.\mathbf{n}$ in order to ensure uniqueness of the potential, if the potential exists. For let ϕ_1 and ϕ_2 be two solutions of (2.1.1), different if this is possible, which have the same specified values on S_1, and for which the values of $\mathbf{w}.\mathbf{n}$ are specified on S_2, and which are both $O(1/R)$ as $R \to \infty$ if V is unbounded. Then $\phi_1 - \phi_2$ satisfies all the conditions satisfied by ϕ above, and hence $\phi_1 - \phi_2 = 0$ everywhere in V, so the two solutions ϕ_1 and ϕ_2 must be identical.

2.5. Sources and sinks

In the first half of this chapter it has been shown that solutions for subsonic irrotational flow can be derived from corresponding incompressible flows, but clearly it is possible to obtain solutions directly. One method of doing this is to consider superpositions of singular solutions, as is sometimes done for incompressible flow, when solutions are built up from sources, doublets, multipoles, vortex pairs, etc.

By analogy with the corresponding solution of Laplace's equation, the singularity at the origin arising from the perturbation potential

$$-(\beta^2x^2+\beta^2y^2+z^2)^{-\frac{1}{2}} \tag{2.5.1}$$

is called a 'subsonic source'. Subsonic sinks have opposite sign, and the higher order multipole solutions can be constructed by differentiation just as for incompressible flow. The velocity is infinite at the singularities, so they must arise (as mathematical devices) inside bodies.

The volume flux† from this subsonic source through any surface S which encloses it is

$$\int_S \mathbf{w}.\mathbf{n}\,dS = \int_S \frac{\beta^2(x\mathbf{i}+y\mathbf{j}+z\mathbf{k}).\mathbf{n}}{(\beta^2x^2+\beta^2y^2+z^2)^{\frac{3}{2}}}\,dS. \tag{2.5.2}$$

Since $\nabla.\mathbf{w}=0$ except at the origin, the form of the surface S is arbitrary and can be taken as a sphere of radius R, centre the origin. On transforming to spherical polar co-ordinates R, ϑ, φ, with the z-axis as polar axis, the volume flux (2.5.2) is

$$\int_0^\pi\int_0^{2\pi} \frac{\beta^2R^3\sin\vartheta\,d\vartheta\,d\varphi}{R^3(\beta^2+M^2\cos^2\vartheta)^{\frac{3}{2}}} = -2\pi\left[\frac{\cos\vartheta}{(\beta^2+M^2\cos^2\vartheta)^{\frac{1}{2}}}\right]_0^\pi = 4\pi. \tag{2.5.3}$$

† For a continuous distribution of sources, $\rho_0 Q$ is the rate of creation of mass per unit volume; hence in linearized theory, Q is the rate of creation of *volume* of fluid per unit volume.

Thus the potential of a subsonic source of unit volume flux is

$$-\frac{1}{4\pi\sqrt{(\beta^2x^2+\beta^2y^2+z^2)}}. \tag{2.5.4}$$

2.6. The general solution of the potential equation

The equation for the potential, ϕ or ϕ', in subsonic flow can be integrated directly in a manner analogous to that used for Laplace's or Poisson's equation, by using an analogue of Green's theorem. The required extension of Green's theorem is obtained as follows.

Let ϕ_1, $\mathbf{v}_1$, $\mathbf{w}_1$ refer to some subsonic flow, and ϕ_2, $\mathbf{v}_2$, $\mathbf{w}_2$ refer to another different flow with the same Mach number. Then an application of the divergence theorem to the vector $\phi_1\mathbf{w}_2$ gives

$$\int_S \phi_1\mathbf{w}_2.\mathbf{n}\,\mathrm{d}S=\int_V (\nabla\phi_1.\mathbf{w}_2+\phi_1\nabla.\mathbf{w}_2)\,\mathrm{d}V, \tag{2.6.1}$$

where $\mathbf{n}$ is the outward normal to the closed surface S. On interchanging the suffices 1 and 2, subtracting the resulting equation from (2.6.1), and remembering that

$$\nabla\phi_1.\mathbf{w}_2=\mathbf{v}_1.\mathbf{w}_2=\mathbf{v}_2.\mathbf{w}_1=\nabla\phi_2.\mathbf{w}_1, \tag{2.6.2}$$

the required analogue of Green's theorem is obtained, namely,

$$\int_S (\phi_1\mathbf{w}_2-\phi_2\mathbf{w}_1).\mathbf{n}\,\mathrm{d}S=\int_V (\phi_1\nabla.\mathbf{w}_2-\phi_2\nabla.\mathbf{w}_1)\,\mathrm{d}V. \tag{2.6.3}$$

Now take ϕ_1, $\mathbf{v}_1$, $\mathbf{w}_1$ to be the solution of the equations

$$\mathbf{v}=\nabla\phi,\quad \mathbf{w}=\mathbf{\Psi}.\mathbf{v},\quad \nabla.\mathbf{w}=Q, \tag{2.6.4}$$

at the point (x_1, y_1, z_1) inside S, and take

$$\phi_2=\frac{1}{\sqrt{\{\beta^2(x-x_1)^2+\beta^2(y-y_1)^2+(z-z_1)^2\}}}=\frac{1}{R_\beta}>0 \quad \text{say,} \tag{2.6.5}$$

where (x, y, z) are the co-ordinates of the elements $\mathrm{d}S$, $\mathrm{d}V$, of surface and volume in (2.6.3). The points (x_1, y_1, z_1) and (x, y, z) are represented by their position vectors $\mathbf{R}_1$ and $\mathbf{R}$ respectively in what follows.

With these values, the integrands in (2.6.3) are singular at $\mathbf{R}_1$, so this point must be excluded from the domain V by a small sphere, S_1, of radius ϵ, centre $\mathbf{R}_1$, enclosing a volume V_1. The surface integral over S_1 is

$$-\int_{S_1}\frac{\beta^2\phi(\mathbf{R})(\mathbf{R}-\mathbf{R}_1).\mathbf{n}}{R_\beta^3}\,\mathrm{d}S-\int_{S_1}\frac{\mathbf{w}(\mathbf{R}).\mathbf{n}}{R_\beta}\,\mathrm{d}S, \tag{2.6.6}$$

where $\mathbf{n}$ is the *inward* normal to the sphere S_1, and $\phi(\mathbf{R})$, $\mathbf{w}(\mathbf{R})$ denote the values of ϕ and $\mathbf{w}$ at the point (x, y, z). On introducing spherical polar co-ordinates as in § 2.5, the integrals in (2.6.6) are easily shown to be

$$4\pi\bar{\phi} - \frac{4\pi\epsilon}{M}\sinh^{-1}\left(\frac{M}{\beta}\right)\overline{\mathbf{w}.\mathbf{n}}, \tag{2.6.7}$$

where the bars denote mean values over S_1. In the limit as $\epsilon \to 0$, this becomes

$$4\pi\phi(\mathbf{R}_1). \tag{2.6.8}$$

The volume integral is

$$-\int_{V-V_1} \frac{Q(\mathbf{R})}{R_\beta}\,\mathrm{d}V, \tag{2.6.9}$$

since $\nabla.\mathbf{w}_2 = 0$ except at $\mathbf{R}_1$. If Q is bounded, this integral exists in the limit as $\epsilon \to 0$.

By combining these results, (2.6.3) gives

$$\phi(\mathbf{R}_1) = \frac{1}{4\pi}\int_S \frac{\mathbf{w}(\mathbf{R}).\mathbf{n}}{R_\beta}\,\mathrm{d}S + \frac{\beta^2}{4\pi}\int_S \frac{\phi(\mathbf{R})(\mathbf{R}-\mathbf{R}_1).\mathbf{n}}{R_\beta^3}\,\mathrm{d}S - \frac{1}{4\pi}\int_V \frac{Q(\mathbf{R})}{R_\beta}\,\mathrm{d}V. \tag{2.6.10}$$

This result enables the value of ϕ to be calculated inside a closed surface S when the values of ϕ and $\mathbf{w}.\mathbf{n}$, or ϕ and $\partial\phi/\partial n$, are known on S, but, just as for the corresponding result for Poisson's equation, the values of ϕ and $\mathbf{w}.\mathbf{n}$ cannot be specified independently on S. In fact, as the uniqueness theorem of § 2.4 shows, a knowledge of either is sufficient to determine the other. The calculation of ϕ in this case is only possible when suitable Green's functions for S are known.

2.7. The general solution of the vector equations

The first-order vector equations,

$$\nabla \wedge \mathbf{v} = \boldsymbol{\zeta}, \quad \nabla.\mathbf{w} = Q, \quad \mathbf{w} = \boldsymbol{\Psi}.\mathbf{v}, \tag{2.7.1}$$

can be integrated directly (Ward, 1952*a*, *b*) by using the integral identity

$$\int_S (\mathbf{v}_1\mathbf{w}_2.\mathbf{n} + \mathbf{v}_2\mathbf{w}_1.\mathbf{n} - \mathbf{v}_1.\mathbf{w}_2\mathbf{n})\,\mathrm{d}S = \int_V \{\mathbf{v}_1\nabla.\mathbf{w}_2 + \mathbf{v}_2\nabla.\mathbf{w}_1 - \mathbf{w}_1\wedge(\nabla\wedge\mathbf{v}_2) - \mathbf{w}_2\wedge(\nabla\wedge\mathbf{v}_1)\}\,\mathrm{d}V, \tag{2.7.2}$$

which is established in Appendix 1.

In this identity, take $\mathbf{v}_1$, $\mathbf{w}_1$ to be the solution of (2.7.1) at the point $\mathbf{R}_1$ inside S, and take $\mathbf{v}_2=\nabla(1/R_\beta)$, etc. Then $\nabla\wedge\mathbf{v}_2=0$ and $\nabla.\mathbf{w}_2=0$ except at $\mathbf{R}_1$, which is to be excluded from the domain of integration by a sphere S_1 of small radius ϵ, as in § 2.6.

The surface integral over S_1 is, since $\mathbf{v}_1.\mathbf{w}_2=\mathbf{v}_2.\mathbf{w}_1$,

$$\bar{\mathbf{v}}\int_{S_1}\mathbf{w}_2.\mathbf{n}\,dS+\bar{\mathbf{w}}\wedge\int_{S_1}\mathbf{v}_2\wedge\mathbf{n}\,dS, \tag{2.7.3}$$

where the bars denote mean values over S_1, and $\mathbf{n}$ is the *inward* normal. The first integral in (2.7.3) has the value 4π, as has been shown in § 2.5. For the second integral,

$$\mathbf{v}_2\wedge\mathbf{n}=\mathbf{w}_2\wedge\mathbf{n}+M^2(z-z_1)\,\mathbf{k}\wedge\mathbf{n}/R_\beta^3, \tag{2.7.4}$$

and $\mathbf{w}_2\wedge\mathbf{n}=0$ on S_1, since $\mathbf{w}_2$ and $\mathbf{n}$ are parallel vectors. On transforming to spherical polar co-ordinates as before,

$$\int_{S_1}\frac{(z-z_1)\,\mathbf{k}\wedge\mathbf{n}}{R_\beta^3}\,dS$$
$$=\int_0^\pi\int_0^{2\pi}\frac{\cos\vartheta(\sin\vartheta\sin\varphi\mathbf{i}-\sin\vartheta\cos\varphi\mathbf{j})\sin\vartheta\,d\vartheta\,d\varphi}{(\beta^2+M^2\cos^2\vartheta)^{\frac{3}{2}}}=0. \tag{2.7.5}$$

Hence, in the limit as $\epsilon\to 0$, the contribution from S_1 is

$$4\pi\mathbf{v}(\mathbf{R}_1). \tag{2.7.6}$$

The surface integral over S can be written

$$\int_S\{(\mathbf{n}.\mathbf{w}_1)\,\mathbf{v}_2+(\mathbf{n}\wedge\mathbf{v}_1)\wedge\mathbf{w}_2\}\,dS, \tag{2.7.7}$$

and if $|\boldsymbol{\zeta}|$ and Q are bounded, the volume integral exists in the limit. Then, on combining the above results, the solution for $\mathbf{v}$ is

$$\begin{aligned}\mathbf{v}(\mathbf{R}_1)= &-\frac{1}{4\pi}\int_S\mathbf{n}.\mathbf{w}(\mathbf{R})\,\nabla\left(\frac{1}{R_\beta}\right)dS\\ &+\frac{\beta^2}{4\pi}\int_S\frac{\{\mathbf{n}\wedge\mathbf{v}(\mathbf{R})\}\wedge(\mathbf{R}-\mathbf{R}_1)}{R_\beta^3}\,dS+\frac{1}{4\pi}\int_V Q(\mathbf{R})\,\nabla\left(\frac{1}{R_\beta}\right)dV\\ &+\frac{\beta^2}{4\pi}\int_V\frac{(\mathbf{R}-\mathbf{R}_1)\wedge\boldsymbol{\zeta}(\mathbf{R})}{R_\beta^3}\,dV.\end{aligned} \tag{2.7.8}$$

This result is a relation between $\mathbf{v}$ inside S and the values of $\mathbf{n}.\mathbf{w}$ and $\mathbf{n}\wedge\mathbf{v}$ on S, but these two quantities cannot be specified independently on S (cf. § 2.6). The surface integrals can be interpreted as the respective contributions of a source distribution of surface density $-\mathbf{n}.\mathbf{w}$, and a vorticity distribution of surface density $-\mathbf{n}\wedge\mathbf{v}$ on S, which represent the velocity field outside S.

It is easily verified that, inside the domain V, the surface integrals satisfy (2.7.1) when both $Q=0$ and $\boldsymbol{\zeta}=0$, and that the volume integrals satisfy these equations when $\boldsymbol{\zeta}=0$ and $Q=0$ respectively.

2.8. Surface and line distributions of vorticity

The velocity field given by

$$\mathbf{v}(\mathbf{R}_1)=\frac{\beta^2}{4\pi}\int_V \frac{(\mathbf{R}-\mathbf{R}_1)\wedge\boldsymbol{\zeta}(\mathbf{R})}{R_\beta^3}\,dV, \tag{2.8.1}$$

where V is the whole of space, satisfies the equations of motion when there are no sources, provided that the integral converges. If the vorticity is confined to a thin layer† of thickness δ, and if dS is an element of one of the bounding surfaces, S say, then

$$\boldsymbol{\zeta}\,dV=\boldsymbol{\zeta}\delta\,dS. \tag{2.8.2}$$

Now if $|\boldsymbol{\zeta}|$ is allowed to become infinite and δ becomes zero in such a way that

$$\lim \delta\boldsymbol{\zeta}=\boldsymbol{\omega}, \tag{2.8.3}$$

where $|\boldsymbol{\omega}|$ is finite, then $\boldsymbol{\omega}$ is the surface density of vorticity on the vortex sheet S, and

$$\mathbf{v}(\mathbf{R}_1)=\frac{\beta^2}{4\pi}\int_S \frac{(\mathbf{R}-\mathbf{R}_1)\wedge\boldsymbol{\omega}(\mathbf{R})}{R_\beta^3}\,dS, \tag{2.8.4}$$

if this integral converges. The surface S must be a *fixed* surface in steady motion, and in general is a cylindrical surface whose generators are parallel to $\mathbf{U}$.

The velocity is discontinuous at a vortex sheet, as can be shown from (2.8.4). If $\mathbf{v}_1$ and $\mathbf{v}_2$ denote the velocities on opposite sides of S at the same point, and if $\mathbf{n}$ is the normal to S at this point, directed from side 1 to side 2, then

$$\mathbf{v}_1-\mathbf{v}_2=\mathbf{n}\wedge\boldsymbol{\omega}, \tag{2.8.5}$$

so there is a discontinuity $\mathbf{n}\wedge\boldsymbol{\omega}$ in $\mathbf{v}$. Since $\boldsymbol{\omega}$ must be tangential to the vortex sheet, (2.8.5) shows that $\mathbf{v}_1-\mathbf{v}_2$ must also be tangential to it, and perpendicular to $\boldsymbol{\omega}$. Hence, in terms of the discontinuity in $\mathbf{v}$, the surface vorticity is given by

$$\boldsymbol{\omega}=\mathbf{n}\wedge(\mathbf{v}_2-\mathbf{v}_1). \tag{2.8.6}$$

† For free vorticity this layer must be bounded by cylindrical surfaces whose generators are parallel to $\mathbf{U}$, since the vorticity must satisfy $\mathbf{U}.\nabla\boldsymbol{\zeta}=0$ ((1.10.8) for steady flow).

If the vorticity is confined to a tube of small cross-sectional area, δ, say, and if ds is an element of the tube axis, then

$$\zeta \mathrm{d}V = |\zeta|\,\delta \mathbf{ds}. \tag{2.8.7}$$

Now let $|\zeta| \to \infty$ and $\delta \to 0$ so that

$$\lim |\zeta|\,\delta = K, \tag{2.8.8}$$

where K must be a constant, then the velocity field of this vortex line of strength K is

$$\mathbf{v} = \frac{\beta^2 K}{4\pi} \int_L \frac{(\mathbf{R} - \mathbf{R}_1) \wedge \mathbf{ds}}{R_\beta^3}, \tag{2.8.9}$$

where the integration is along the vortex line, L. A free vortex line must be a straight line parallel to $\mathbf{U}$, and it can be shown that the integral in (2.8.9) converges, even if L is infinite, except on L.

2.9. Horseshoe vortices

The most commonly considered vortex system is the simple horseshoe vortex. This consists of a straight bound vortex line of finite length, perpendicular to $\mathbf{U}$, with two semi-infinite free vortex lines stretching downstream and parallel to $\mathbf{U}$ from the ends of the bound vortex. The limiting case when the separation of the two free vortices tends to zero, and the strength K tends to infinity in such a way that their product remains finite, is called a vortex pair.

The velocity field of a horseshoe vortex is easily calculated from (2.8.9), or it can be derived by the method of § 2.3 from the results for incompressible flow.† In the latter case the circulations are required to be the same, so $C = \beta$; also, if $k = 1$, the dimensions perpendicular to the stream direction remain the same, and then the horseshoe vortex has the same configuration in the related flows.

If the bound vortex lies along the x-axis for $-a \leqslant x \leqslant a$ with positive circulation K, and the two free trailing vortices lie parallel to the z-axis in $z > 0$, then the components of the perturbation velocity field, $\mathbf{v}$, are

$$u = \frac{K}{4\pi}\left\{\frac{y}{(x+a)^2+y^2}\left[1+\frac{z}{\sqrt{[\beta^2(x+a)^2+\beta^2y^2+z^2]}}\right] - \frac{y}{(x-a)^2+y^2}\left[1+\frac{z}{\sqrt{[\beta^2(x-a)^2+\beta^2y^2+z^2]}}\right]\right\}, \tag{2.9.1}$$

† See, for example, H. Glauert, *The Elements of Aerofoil and Airscrew Theory* (Cambridge, 1937), chap. XII.

$$v=\frac{K}{4\pi}\left\{\frac{x-a}{(x-a)^2+y^2}\left[1+\frac{z}{\sqrt{[\beta^2(x-a)^2+\beta^2y^2+z^2]}}\right]\right.$$

$$-\frac{x+a}{(x+a)^2+y^2}\left[1+\frac{z}{\sqrt{[\beta^2(x+a)^2+\beta^2y^2+z^2]}}\right]$$

$$\left.+\frac{\beta^2z^2}{\beta^2y^2+z^2}\left[\frac{x-a}{\sqrt{[\beta^2(x-a)^2+\beta^2y^2+z^2]}}-\frac{x+a}{\sqrt{[\beta^2(x+a)^2+\beta^2y^2+z^2]}}\right]\right\}, \tag{2.9.2}$$

$$w=\frac{K\beta^2}{4\pi}\frac{y}{\beta^2y^2+z^2}$$

$$\times\left[\frac{x+a}{\sqrt{[\beta^2(x+a)^2+\beta^2y^2+z^2]}}-\frac{x-a}{\sqrt{[\beta^2(x-a)^2+\beta^2y^2+z^2]}}\right]. \tag{2.9.3}$$

2.10. Surface distributions of sources

The potential and velocity fields for a distribution of sources alone are given from (2.6.10) and (2.7.8) respectively as

$$\phi(\mathbf{R}_1)=-\frac{1}{4\pi}\int_V\frac{Q(\mathbf{R})}{R_\beta}\,dV, \tag{2.10.1}$$

$$\mathbf{v}(\mathbf{R}_1)=\frac{1}{4\pi}\int_V Q(\mathbf{R})\,\nabla\left(\frac{1}{R_\beta}\right)dV, \tag{2.10.2}$$

where V is the whole of space for which $Q\neq 0$, if these integrals converge.

For a distribution of sources of surface density q on a surface S,

$$\phi(\mathbf{R}_1)=-\frac{1}{4\pi}\int_S\frac{q(\mathbf{R})}{R_\beta}\,dS, \tag{2.10.3}$$

$$\mathbf{v}(\mathbf{R}_1)=\frac{1}{4\pi}\int_S q(\mathbf{R})\,\nabla\left(\frac{1}{R_\beta}\right)dS. \tag{2.10.4}$$

The potential is continuous, and the velocity is discontinuous at such a source layer. If $\mathbf{v}_1$, $\mathbf{v}_2$ and $\mathbf{n}$ have the meanings given to them in § 2.8, then it follows from (2.10.3) and (2.10.4), or directly from (2.7.1), that

$$\mathbf{n}\wedge(\mathbf{v}_2-\mathbf{v}_1)=0 \quad\text{and}\quad \mathbf{n}.(\mathbf{w}_2-\mathbf{w}_1)=q; \tag{2.10.5}$$

whence

$$\mathbf{v}_2-\mathbf{v}_1=\frac{q\mathbf{n}}{\mathbf{n}.\mathbf{\Psi}.\mathbf{n}} \quad\text{and}\quad \mathbf{w}_2-\mathbf{w}_1=\frac{q\mathbf{\Psi}.\mathbf{n}}{\mathbf{n}.\mathbf{\Psi}.\mathbf{n}}. \tag{2.10.6}$$

If the layer is parallel to $\mathbf{U}$ at every point (as is usually the case) and $\boldsymbol{\nu}$ is the normal, then, since $\mathbf{U}.\boldsymbol{\nu}=0$,

$$\boldsymbol{\nu}.\mathbf{\Psi}.\boldsymbol{\nu}=1, \tag{2.10.7}$$

and the discontinuities in $\mathbf{v}$ and $\mathbf{w}$ are

$$\mathbf{v}_2-\mathbf{v}_1=\mathbf{w}_2-\mathbf{w}_1=q\boldsymbol{\nu}. \tag{2.10.8}$$

CHAPTER 3

THE GENERAL SOLUTIONS OF THE LINEARIZED EQUATIONS FOR SUPERSONIC FLOW

3.1. The characteristics of the potential equation

In linearized theory, the acoustic speed, c_0, is constant for the whole fluid, so if a small disturbance is introduced momentarily at some point (x_1, y_1, z_1), then the wave front of the resulting perturbations in the stream is a sphere whose radius is increasing at a rate c_0, and which is being swept downstream bodily with velocity U. When the stream is supersonic ($U/c_0 = M > 1$), the wave fronts at successive times are enveloped by a half-cone of semi-angle $\mu = \operatorname{cosec}^{-1} M = \cot^{-1} \sqrt{(M^2 - 1)}$, whose equation is

$$z - z_1 = \cot \mu \sqrt{[(x - x_1)^2 + (y - y_1)^2]} > 0. \tag{3.1.1}$$

For steady flow, any small disturbance introduced must also be steady, and it is easily seen that the effects of such a disturbance must be confined to the interior of the half-cone (3.1.1), there being no disturbance of the fluid outside it. This is in direct contrast with the state of affairs for subsonic flow, in which the effects of any steady disturbance are felt throughout the fluid. The above considerations can be generalized immediately to cover the case of a distribution of disturbances introduced into a supersonic stream. The perturbations are then confined to the downstream side of the envelope of all half-cones of the type (3.1.1) whose vertices coincide with the points of disturbance.

The angle μ is called the Mach angle, and cones of the type (3.1.1) are called Mach cones, or characteristic cones. Envelopes of characteristic cones, or the cones themselves, are characteristic surfaces, and they play a very important part in the study of the equations of supersonic flow.

The linearized potential equation for supersonic flow can be written

$$\frac{\partial^2 \phi}{\partial x^2} + \frac{\partial^2 \phi}{\partial y^2} - B^2 \frac{\partial^2 \phi}{\partial z^2} = 0, \tag{3.1.2}$$

where $$B = \sqrt{(M^2 - 1)} = \cot \mu > 0,$$

and is an equation of hyperbolic type. Unlike elliptic equations, hyperbolic equations admit of solutions that are piecewise analytic. The investigation of surfaces of discontinuity given in § 1.11 shows that discontinuities in the normal component of velocity and its derivatives are allowable in linearized theory on characteristic surfaces, $G(x, y, z) = 0$ such that

$$\left(\frac{\partial G}{\partial x}\right)^2 + \left(\frac{\partial G}{\partial y}\right)^2 - B^2\left(\frac{\partial G}{\partial z}\right)^2 = 0 \quad \text{on} \quad G = 0. \tag{3.1.3}$$

It is easily verified that this relation is satisfied by (3.1.1) for the characteristic cone, and hence for any envelope of such cones; this shows that the mathematical and physical definitions of characteristic surfaces for (3.1.2) are equivalent. However, the physical definition distinguishes between the upstream and downstream halves of characteristic cones, which the mathematical definition does not.

3.2. The initial value problem: uniqueness

For source-free irrotational flow, the z-component of the identity (A. 1.2), established in Appendix 1, is

$$\int_S (\mathbf{k}.\mathbf{vw}.\mathbf{n} - \tfrac{1}{2}\mathbf{v}.\mathbf{wk}.\mathbf{n})\,\mathrm{d}S = 0, \tag{3.2.1}$$

and in terms of the perturbation potential, ϕ, this can be written

$$\int_S \left\{2\frac{\partial\phi}{\partial z}\left(l\frac{\partial\phi}{\partial x} + m\frac{\partial\phi}{\partial y}\right) - n\left[\left(\frac{\partial\phi}{\partial x}\right)^2 + \left(\frac{\partial\phi}{\partial y}\right)^2 + B^2\left(\frac{\partial\phi}{\partial z}\right)^2\right]\right\}\mathrm{d}S = 0, \tag{3.2.2}$$

where l, m, n are the direction cosines of the outward normal at any point of S.

Now let a part of S, S_1 say, be a surface whose equation is

$$z = f(x, y), \tag{3.2.3}$$

where $\partial f/\partial x$ and $\partial f/\partial y$ are continuous on S_1. Then, on S_1,

$$\frac{l}{n} = -\frac{\partial f}{\partial x}, \quad \frac{m}{n} = -\frac{\partial f}{\partial y}. \tag{3.2.4}$$

If the value of ϕ on S_1, $\phi(x, y, f)$, is denoted by $[\phi]$, and likewise for other quantities, then

$$\frac{\partial}{\partial x}[\phi] = \left[\frac{\partial\phi}{\partial x}\right] - \frac{l}{n}\left[\frac{\partial\phi}{\partial z}\right], \quad \frac{\partial}{\partial y}[\phi] = \left[\frac{\partial\phi}{\partial y}\right] - \frac{m}{n}\left[\frac{\partial\phi}{\partial z}\right], \tag{3.2.5}$$

and the contribution to (3.2.2) for S_1 can be written

$$-\int_{S_1}\left[\left(\frac{\partial\phi}{\partial x}-\frac{l}{n}\frac{\partial\phi}{\partial z}\right)^2+\left(\frac{\partial\phi}{\partial y}-\frac{m}{n}\frac{\partial\phi}{\partial z}\right)^2+\left(B^2-\frac{l^2+m^2}{n^2}\right)\left(\frac{\partial\phi}{\partial z}\right)^2\right]n\,dS$$

$$=-\operatorname{sgn}(n)\iint_{S_1}\left\{\left(\frac{\partial}{\partial x}[\phi]\right)^2+\left(\frac{\partial}{\partial y}[\phi]\right)^2+\left(B^2-\left(\frac{\partial f}{\partial x}\right)^2-\left(\frac{\partial f}{\partial y}\right)^2\right)\left[\frac{\partial\phi}{\partial z}\right]^2\right\}dx\,dy. \quad (3.2.6)$$

It follows from (3.1.3) and (3.2.3) that, when S_1 is a characteristic surface, the last term in the integrand in (3.2.6) vanishes, since

$$B^2-\left(\frac{\partial f}{\partial x}\right)^2-\left(\frac{\partial f}{\partial y}\right)^2=0. \quad (3.2.7)$$

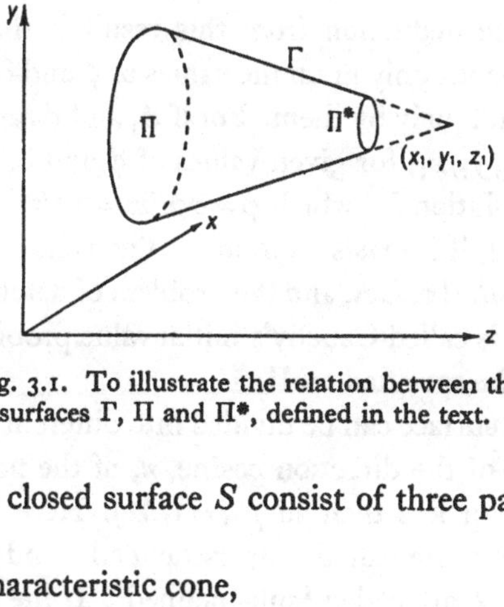

Fig. 3.1. To illustrate the relation between the surfaces Γ, Π and Π*, defined in the text.

Next, let the closed surface S consist of three parts, as follows (see fig. 3.1):

(i) Γ, the characteristic cone,

$$z=z_1-B\sqrt{[(x-x_1)^2+(y-y_1)^2]} \quad (0\leqslant z\leqslant z_0<z_1),$$

(ii) Π, the plane,

$$z=0, \quad B^2[(x-x_1)^2+(y-y_1)^2]\leqslant z_1^2,$$

(iii) Π*, the plane,

$$z=z_0<z_1, \quad B^2[(x-x_1)^2+(y-y_1)^2]\leqslant(z_1-z_0)^2,$$

and let

$$\phi=0, \quad \frac{\partial\phi}{\partial z}=0 \quad (3.2.8)$$

on Π. Then, since $n>0$ on Γ, $l=m=0$, $n=1$ on Π*, the identity (3.2.2), combined with (3.2.6) for the characteristic surface Γ, gives

$$\iint_\Gamma \left\{\left(\frac{\partial}{\partial x}[\phi]\right)^2 + \left(\frac{\partial}{\partial y}[\phi]\right)^2\right\} dx\,dy + \iint_{\Pi^*} \left\{\left(\frac{\partial \phi}{\partial x}\right)^2 + \left(\frac{\partial \phi}{\partial y}\right)^2 + B^2\left(\frac{\partial \phi}{\partial z}\right)^2\right\} dx\,dy = 0. \quad (3.2.9)$$

Hence on Π^*

$$\frac{\partial \phi}{\partial x} = \frac{\partial \phi}{\partial y} = \frac{\partial \phi}{\partial z} = 0, \quad (3.2.10)$$

and on Γ

$$\frac{\partial}{\partial x}[\phi] = \frac{\partial}{\partial y}[\phi] = 0. \quad (3.2.11)$$

By allowing z_0 to vary from 0 to z, the above argument shows that (3.2.10) is true for all points in the interior of Γ, and in particular at the vertex (x_1, y_1, z_1), and hence that $\phi = 0$ in this domain, since $\phi = 0$ on Π.

An immediate deduction from this result is that the value of $\phi(x_1, y_1, z_1)$ depends only upon the values of ϕ and $\partial\phi/\partial z$ on Π, and is determined uniquely by them. For if ϕ_1 and ϕ_2 are two solutions of (3.1.2) at (x_1, y_1, z_1) for given values of ϕ and $\partial\phi/\partial z$ on Π, then $\phi_1 - \phi_2$ is the solution for which $\phi = \partial\phi/\partial z = 0$ on Π, so $\phi_1 - \phi_2 = 0$, and the solution, if it exists, is unique. The values of ϕ and $\partial\phi/\partial z$ are called the *initial values*, and the problem of determining ϕ when these are given is called Cauchy's initial value problem; *both* initial values have to be specified on Π.

In general, a surface can be divided into different parts for which the magnitude of the direction cosine, n, of the normal is greater than, equal to, or less than $\sin\mu$ $(=1/M)$. Here, those parts for which $|n| > \sin\mu$ are called 'super-inclined', and those parts for which $|n| < \sin\mu$ are called 'sub-inclined'. If the equation of the surface is $g(x, y, z) = 0$, then

$$|n| = \left| \frac{\dfrac{\partial g}{\partial z}}{\sqrt{\left[\left(\dfrac{\partial g}{\partial x}\right)^2 + \left(\dfrac{\partial g}{\partial y}\right)^2 + \left(\dfrac{\partial g}{\partial z}\right)^2\right]}} \right| \gtrless \sin\mu, \quad (3.2.12)$$

for super-inclined and sub-inclined surfaces respectively. This gives

$$\left(\frac{\partial g}{\partial x}\right)^2 + \left(\frac{\partial g}{\partial y}\right)^2 - B^2\left(\frac{\partial g}{\partial z}\right)^2 \lessgtr 0, \quad (3.2.13)$$

which provides an alternative definition, and shows incidentally that $|n| = \sin\mu$ for characteristic surfaces (cf. (3.1.3)).

The proof of uniqueness given above can be generalized easily for the case where Π is replaced by any super-inclined surface, showing that the specification of ϕ and $\partial\phi/\partial z$, or ϕ and $\partial\phi/\partial n$ on any such surface leads to a unique solution for ϕ. Hadamard (1923) calls Cauchy's problem in this case 'properly set', because both ϕ and $\partial\phi/\partial n$ are required for the determination of ϕ.

The proof can also be generalized in the same way when Π is replaced by a partly sub-inclined surface, but in this case, Cauchy's problem is not properly set, and ϕ and $\partial\phi/\partial n$ cannot be specified independently on sub-inclined surfaces. A uniqueness result when $\partial\phi/\partial n$ is specified is proved in §4.9.

3.3. Dependence and influence domains

The set of points (x, y, z) such that

$$z \leqslant z_1 - B\sqrt{[(x-x_1)^2+(y-y_1)^2]} \tag{3.3.1}$$

which lie inside the characteristic half-cone, Γ, with vertex at P, (x_1, y_1, z_1), is called the *dependence domain*, $\mathscr{D}_P$, of P (see fig. 3.2), because the value of ϕ at P depends only upon the values of ϕ in $\mathscr{D}_P$, as shown in §3.2.

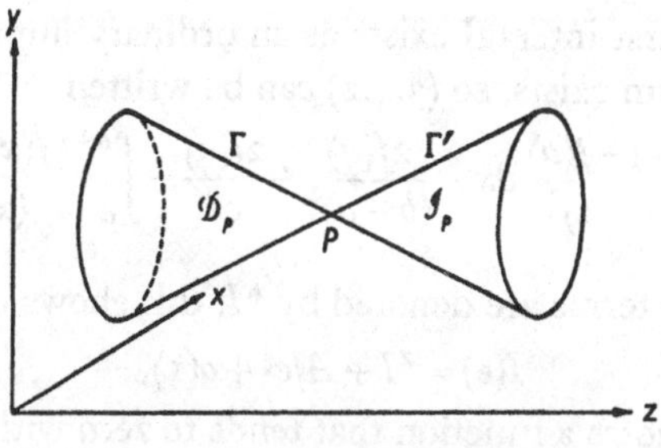

Fig. 3.2. Dependence and influence domains for the point P.

If Q is another point, then it follows that the value of ϕ at Q can be affected by changes in ϕ at P, only if P lies in the dependence domain $\mathscr{D}_Q$ of Q. The set of points Q such that this is so is called the *influence domain*, $\mathscr{I}_P$, of P. If Q is taken to be the point (x, y, z), then it follows from (3.3.1) that P is a point of $\mathscr{D}_Q$ if

$$z_1 \leqslant z - B\sqrt{[(x-x_1)^2+(y-y_1)^2]}; \tag{3.3.2}$$

hence $\mathscr{I}_P$ is defined by

$$z \geqslant z_1 + B\sqrt{[(x-x_1)^2+(y-y_1)^2]}, \tag{3.3.3}$$

and is the interior of the half-cone Γ', such that Γ and Γ' together make up the whole characteristic cone (see fig. 3.2)

$$B^2[(x-x_1)^2+(y-y_1)^2] = (z-z_1)^2. \tag{3.3.4}$$

3.4. The 'finite part' of a divergent integral

Before going on to consider the general solution of the potential equation, it is necessary to define a concept in integration which enables certain types of divergent integrals to be handled.

It is simplest to begin by considering the special case of the integral

$$I(\epsilon) = \int_{a+\epsilon}^{b} \frac{f(x)}{(x-a)^{\frac{3}{2}}}\,dx, \tag{3.4.1}$$

where $f'(x)$ is assumed to exist and to be bounded for $a \leqslant x \leqslant b$. If $\epsilon > 0$, then this integral exists, but if $\epsilon = 0$, then the integral is divergent, and no meaning can be assigned to $I(0)$. The integral $I(\epsilon)$ can be written

$$I(\epsilon) = \int_{a+\epsilon}^{b} \frac{f(x)-f(a)}{(x-a)^{\frac{3}{2}}}\,dx + f(a)\int_{a+\epsilon}^{b} \frac{dx}{(x-a)^{\frac{3}{2}}}$$

$$= \int_{a+\epsilon}^{b} \frac{f(x)-f(a)}{(x-a)^{\frac{3}{2}}}\,dx + f(a)\left(\frac{2}{\epsilon^{\frac{1}{2}}} - \frac{2}{(b-a)^{\frac{1}{2}}}\right). \tag{3.4.2}$$

As $\epsilon \to 0$, the first integral exists as an ordinary improper integral, and the last term exists, so (3.4.2) can be written

$$I(\epsilon) = \int_{a}^{b} \frac{f(x)-f(a)}{(x-a)^{\frac{3}{2}}}\,dx - \frac{2f(a)}{(b-a)^{\frac{1}{2}}} + \frac{2f(a)}{\epsilon^{\frac{1}{2}}} - \int_{a}^{a+\epsilon} \frac{f(x)-f(a)}{(x-a)^{\frac{3}{2}}}\,dx. \tag{3.4.3}$$

If the first two terms are denoted by $*I$, this shows that

$$I(\epsilon) = {*I} + A/\epsilon^{\frac{1}{2}} + o(1), \tag{3.4.4}$$

where $o(1)$ denotes a function that tends to zero with ϵ, whence

$$*I = \lim_{\epsilon \to 0}\,(I(\epsilon) - A/\epsilon^{\frac{1}{2}}). \tag{3.4.5}$$

Hadamard (1923) called this unique limit, $*I$, the *finite part* of $I(0)$, and it is denoted by a star placed in front of the integral, thus

$$*\!\int_{a}^{b} \frac{f(x)}{(x-a)^{\frac{3}{2}}}\,dx = \int_{a}^{b} \frac{f(x)-f(a)}{(x-a)^{\frac{3}{2}}}\,dx - \frac{2f(a)}{(b-a)^{\frac{1}{2}}}. \tag{3.4.6}$$

This definition of the finite part of a divergent integral, and its following generalizations, were introduced by Hadamard for the express purpose of solving boundary-value problems for second-order linear hyperbolic differential equations.

Before going on to generalize the definition, the two following important results must be noted. First, if it is known that $I(\epsilon)=0$ for all values of ϵ, then it follows from (3.4.4) that $A=0$, and $*I=o(1)$, which implies that $*I=0$. Thus

$$\textit{if } I(\epsilon)=0 \textit{ for all } \epsilon, \textit{ then } *I=0. \tag{3.4.7}$$

Secondly, consider

$$\frac{\partial}{\partial a}\int_a^b \frac{f(x)}{(x-a)^{\frac{1}{2}}}\,dx. \tag{3.4.8}$$

By differentiating formally this becomes

$$\frac{1}{2}\int_a^b \frac{f(x)}{(x-a)^{\frac{3}{2}}}\,dx-\left[\frac{f(x)}{(x-a)^{\frac{1}{2}}}\right]_{x=a},$$

which is meaningless as written. In order to evaluate (3.4.8), it can be written

$$\frac{\partial}{\partial a}\left\{\int_a^b \frac{f(x)-f(a)}{(x-a)^{\frac{1}{2}}}\,dx+f(a)\int_a^b \frac{dx}{(x-a)^{\frac{1}{2}}}\right\}, \tag{3.4.9}$$

and on carrying out the differentiation, this gives

$$\frac{1}{2}\int_a^b \frac{f(x)-f(a)}{(x-a)^{\frac{3}{2}}}\,dx-\frac{f(a)}{(b-a)^{\frac{1}{2}}}, \tag{3.4.10}$$

the terms in $f'(a)$ having cancelled. Comparison with (3.4.6) shows that

$$\frac{\partial}{\partial a}\int_a^b \frac{f(x)}{(x-a)^{\frac{1}{2}}}\,dx=\frac{1}{2}\,{}^*\!\!\int_a^b \frac{f(x)}{(x-a)^{\frac{3}{2}}}\,dx={}^*\!\!\int_a^b \frac{\partial}{\partial a}\left\{\frac{f(x)}{(x-a)^{\frac{1}{2}}}\right\}dx. \tag{3.4.11}$$

Thus to evaluate (3.4.8), change the order of differentiation and integration, and take the finite part.

In a similar way, it can be shown that

$$I(\epsilon)=\int_{a+\epsilon}^b \frac{f(x)}{(x-a)^{N+k}}\,dx=\sum_{n=0}^{N-1}\frac{A_n}{\epsilon^{n+k}}+{}^*I+o(1), \tag{3.4.12}$$

where $N(\geqslant 1)$ is an integer, and $0<k<1$. The coefficients A_n and $*I$ are unique provided that the first N derivatives of $f(x)$ are bounded and continuous in $a\leqslant x\leqslant b$. Just as before, the finite part of $I(\epsilon)$ when $\epsilon=0$ is defined as

$$*I={}^*\!\!\int_a^b \frac{f(x)}{(x-a)^{N+k}}\,dx=\lim_{\epsilon\to 0}\left(I(\epsilon)-\sum_{n=0}^{N-1}\frac{A_n}{\epsilon^{n+k}}\right), \tag{3.4.13}$$

and it has the properties given in (3.4.7) and (3.4.11), as can be verified. When k in (3.4.12) is an integer, logarithmic terms occur and the above definition of the finite part fails to give a unique value. However, in all subsequent applications $k=\frac{1}{2}$, so this difficulty does not arise.

The concept of the finite part can be generalized still further to apply to multiple integrals for use in the subsequent analysis. This was done by Hadamard in his book, but the following method is due to Robinson (1948*b*); further extensions have been made by Lomax, Heaslet and Fuller (1951).

Let $G(x, y, z)$ be an integral function of all three variables, which is real when x, y and z are real; then the equation

$$G(x, y, x)=0 \qquad (3.4.14)$$

determines a surface, Σ, which divides space into regions V_n in which $G(x, y, z)$ is of constant sign. $G(x, y, z)$ is supposed to change sign across any ordinary point of Σ.

Let a scalar or vector function $f(x, y, z)$ be defined by

$$\begin{aligned} f(x,y,z) &= f_{-n}(x,y,z)\,G^{-n+\frac{1}{2}}+f_{-n+1}(x,y,z)\,G^{-n+\frac{3}{2}}+\ldots \\ &\quad +f_{-1}(x,y,z)\,G^{-\frac{1}{2}}+f_0(x,y,z)+f_1(x,y,z)\,G^{\frac{1}{2}}+\ldots+f_m(x,y,z)\,G^{m-\frac{1}{2}}, \end{aligned} \qquad (3.4.15)$$

where m and n are finite positive integers, and where $f_{-n}, \ldots, f_m$ are either all analytic everywhere except on Σ, or have derivatives of a sufficiently high order which are bounded in the neighbourhood of Σ. The analytic expressions for these functions may be different in the different regions V_n.

Let I denote any of the integrals

$$\int_C f\,ds, \quad \int_S f\,dS, \quad \int_V f\,dV, \qquad (3.4.16)$$

where the integrations are over a line C, a surface S, or a volume V; let $C(\epsilon)$, $S(\epsilon)$, $V(\epsilon)$ denote the sets of points in C, S, V respectively for which

$$|G(x, y, z)|>\epsilon>0, \qquad (3.4.17)$$

and let $I(\epsilon)$ denote the integral corresponding to I, taken over $C(\epsilon)$, $S(\epsilon)$ or $V(\epsilon)$. Then $I(\epsilon)$ is finite and of the form

$$I(\epsilon)=\sum_{n=0}^{N}\frac{A_n}{\epsilon^{n+\frac{1}{2}}}+{}^*I+o(1), \qquad (3.4.18)$$

where $*I$ is independent of ϵ, and $o(1)$ denotes a function which tends to zero with ϵ. The finite part, $*I$, of I is then uniquely defined by

$$*I = \lim_{\epsilon \to 0}\left(I(\epsilon) - \sum_{n=0}^{N} \frac{A_n}{\epsilon^{n+\frac{1}{2}}}\right). \tag{3.4.19}$$

The finite part of I has the following properties which are not difficult to verify:

(i) If I converges, then $*I = I$.

(ii) The value of $*I$ is invariant with respect to co-ordinate transformations provided that the transformations are not singular on Σ.

(iii) If the integrands involve vector quantities, then $*I$ is invariant for rotations of the co-ordinate axes.

(iv) If $I(\epsilon)$ vanishes for all values of ϵ, then $*I = 0$.

(v) If $f(x, y, z)$ depends upon some parameter, a, and in particular if G depends also upon a, so that Σ varies as a varies, then $\partial *I/\partial a$ can be evaluated by interchanging the order of integration and differentiation; for example,

$$\frac{\partial}{\partial a} {}^*\!\!\int f\,\mathrm{d}S = {}^*\!\!\int \frac{\partial f}{\partial a}\,\mathrm{d}S.$$

The above definition and properties can be extended to cover functions of any finite number of independent variables, with analogous results.

3.5. Supersonic sources and sinks

Corresponding to the solution for a source in subsonic flow (§ 2.5), a fundamental solution of the potential equation (3.1.2) for supersonic flow is

$$\frac{1}{R_B} = \frac{1}{\sqrt{\{(z-z_1)^2 - B^2(x-x_1)^2 - B^2(y-y_1)^2\}}}. \tag{3.5.1}$$

This solution is real inside the characteristic cone with vertex at (x_1, y_1, z_1) and is singular on this cone. The quantity R_B is sometimes called the hyperbolic distance between (x, y, z) and (x_1, y_1, z_1), and is taken to be positive when real.

Consider now the particular solution

$$\left.\begin{aligned} \phi &= -\frac{1}{R_B} \quad \text{for} \quad R_B^2 > 0,\ z > z_1, \\ &= 0 \text{ at all other points.} \end{aligned}\right\} \tag{3.5.2}$$

The disturbance represented by this solution is zero outside the domain of influence of (x_1, y_1, z_1) and, by analogy with a subsonic source, is said to be that for a supersonic source at (x_1, y_1, z_1). That this is a reasonable name can be seen by calculating the volume flux through a closed surface surrounding the source. The value of $\mathbf{w}$ corresponding to (3.5.2) is

$$\left.\begin{aligned} \mathbf{w} &= -\boldsymbol{\Psi}.\nabla\left(\frac{1}{R_B}\right) = -\frac{B^2(\mathbf{R}-\mathbf{R}_1)}{R_B^3} \quad \text{for} \quad R_B^2 > 0,\ z > z_1, \\ &= 0 \text{ at all other points.} \end{aligned}\right\} \tag{3.5.3}$$

The flux integral
$$\int_S \mathbf{w}.\mathbf{n}\,\mathrm{d}S, \tag{3.5.4}$$

taken over any enclosing surface, diverges, but the finite part can be taken since $\mathbf{w}.\mathbf{n}$ satisfies all the conditions of § 3.4. The volume flux is defined to be
$${}^{*}\!\!\int_S \mathbf{w}.\mathbf{n}\,\mathrm{d}S \tag{3.5.5}$$

in supersonic flow, but it must be shown that this integral is invariant for changes of S, so some care is required in its evaluation. Since $\mathbf{w}$ vanishes outside the influence domain of $\mathbf{R}_1$, only that part of S which lies inside this domain, S_1 say, need be considered.

In spherical polar co-ordinates with origin at $\mathbf{R}_1$ and polar axis parallel to the stream direction, let the vector equation of S_1 be

$$\mathbf{R}-\mathbf{R}_1 = \chi(\vartheta, \varphi)\,\boldsymbol{\alpha}(\vartheta, \varphi), \tag{3.5.6}$$

where $\boldsymbol{\alpha}$ is the unit radial vector. Then for any element $\mathrm{d}S$ of S_1, formally,

$$\begin{aligned} \mathbf{n}\,\mathrm{d}S &= \frac{\partial}{\partial\vartheta}(\chi\boldsymbol{\alpha}) \wedge \frac{\partial}{\partial\varphi}(\chi\boldsymbol{\alpha})\,\mathrm{d}\vartheta\,\mathrm{d}\varphi \\ &= \left\{\chi^2\frac{\partial\boldsymbol{\alpha}}{\partial\vartheta}\wedge\frac{\partial\boldsymbol{\alpha}}{\partial\varphi} + \chi\frac{\partial\chi}{\partial\vartheta}\boldsymbol{\alpha}\wedge\frac{\partial\boldsymbol{\alpha}}{\partial\varphi} + \chi\frac{\partial\chi}{\partial\varphi}\frac{\partial\boldsymbol{\alpha}}{\partial\vartheta}\wedge\boldsymbol{\alpha}\right\}\mathrm{d}\vartheta\,\mathrm{d}\varphi. \end{aligned} \tag{3.5.7}$$

Hence from (3.5.3), (3.5.6) and (3.5.7),

$$\mathbf{w}.\mathbf{n}\,\mathrm{d}S = -\frac{B^2\chi^3}{R_B^3}\,\boldsymbol{\alpha}.\frac{\partial\boldsymbol{\alpha}}{\partial\vartheta}\wedge\frac{\partial\boldsymbol{\alpha}}{\partial\varphi}\,\mathrm{d}\vartheta\,\mathrm{d}\varphi. \tag{3.5.8}$$

If α_1, α_2, α_3 are the rectangular components, or direction cosines, of $\boldsymbol{\alpha}$, then

$$R_B^3 = \chi^3(\alpha_3^2 - B^2\alpha_1^2 - B^2\alpha_2^2)^{\frac{3}{2}}; \tag{3.5.9}$$

hence, finally, the integral (3.5.5) is

$$-B^2 \overset{*}{\iint}_{S_1} \frac{\boldsymbol{\alpha}.\dfrac{\partial\boldsymbol{\alpha}}{\partial\vartheta}\wedge\dfrac{\partial\boldsymbol{\alpha}}{\partial\varphi}}{(\alpha_3^2-B^2\alpha_1^2-B^2\alpha_2^2)^{\frac{3}{2}}}\,d\vartheta\,d\varphi, \tag{3.5.10}$$

which is independent of $\chi(\vartheta, \varphi)$, and so is independent of the particular form of S_1. The integral (3.5.10) is

$$-B^2 \overset{*}{\int_0^{\cos^{-1}(B/M)}} \int_0^{2\pi} \frac{\sin\vartheta}{(M^2\cos^2\vartheta - B^2)^{\frac{3}{2}}}\,d\vartheta\,d\varphi$$

$$= -2\pi B^2 \overset{*}{\int_0^1} \frac{d\kappa}{(M^2\kappa^2-B^2)^{\frac{3}{2}}}. \tag{3.5.11}$$

Now $$-\int_{\sqrt{(B^2+\epsilon)}/M}^{1} \frac{d\kappa}{(M^2\kappa^2-B^2)^{\frac{3}{2}}} = \left[\frac{\kappa}{B^2(M^2\kappa^2-B)^{\frac{1}{2}}}\right]_{\sqrt{(B^2+\epsilon)}/M}^{1}$$

$$= \frac{1}{B^2} - \frac{1}{MB\epsilon^{\frac{1}{2}}} + o(1). \tag{3.5.12}$$

Hence the volume flux is 2π, and the potential

$$\left.\begin{aligned} \phi &= -\frac{1}{2\pi R_B} \quad \text{for} \quad R_B^2 > 0,\ z > z_1, \\ &= 0 \text{ at all other points,} \end{aligned}\right\} \tag{3.5.13}$$

represents a supersonic source of unit volume flux. The corresponding potential with opposite sign represents a supersonic sink, and higher order singularities can be constructed by differentiation as in subsonic flow.

The singularity at $\mathbf{R}_1$ corresponding to the complementary potential

$$\left.\begin{aligned} \phi &= -\frac{1}{2\pi R_B}, \quad R_B^2 > 0,\ z < z_1, \\ &= 0 \text{ at all other points,} \end{aligned}\right\} \tag{3.5.14}$$

has been called a counter-source by Robinson (1948 *b*).

3.6. The general solution of the potential equation

The integration of the potential equation for supersonic flow can be accomplished with the aid of the integral identity

$$\int_S (\phi_1\mathbf{w}_2 - \phi_2\mathbf{w}_1).\mathbf{n}\,dS = \int_V (\phi_1\nabla.\mathbf{w}_2 - \phi_2\nabla.\mathbf{w}_1)\,dV, \tag{3.6.1}$$

which has been established already in § 2.6. The slightly generalized case when there is a source distribution is considered here, so that

ϕ_1, $\mathbf{v}_1$, $\mathbf{w}_1$ in (3.6.1) are taken to be the solution of

$$\mathbf{v}=\nabla\phi, \quad \mathbf{w}=\mathbf{\Psi}.\mathbf{v}, \quad \nabla.\mathbf{w}=Q, \tag{3.6.2}$$

at the point $\mathbf{R}_1$, for given boundary values on some surface, S, which lies upstream from this point (in its domain of dependence).

Take ϕ_2 to be the potential

$$\left.\begin{aligned}\phi_2&=\frac{1}{R_B} \quad \text{for} \quad R_B^2>0,\ z<z_1,\\ &=0 \text{ at all other points,}\end{aligned}\right\} \tag{3.6.3}$$

so that ϕ_2 is the potential of a counter-sink of strength 2π at $\mathbf{R}_1$, and take the volume of integration in (3.6.1) to be bounded by that part of $R_B=0$ $(z<z_1)$, which lies downstream from S, Γ say, plus that part of S which lies inside Γ, as in fig. 3.3. The surface integral

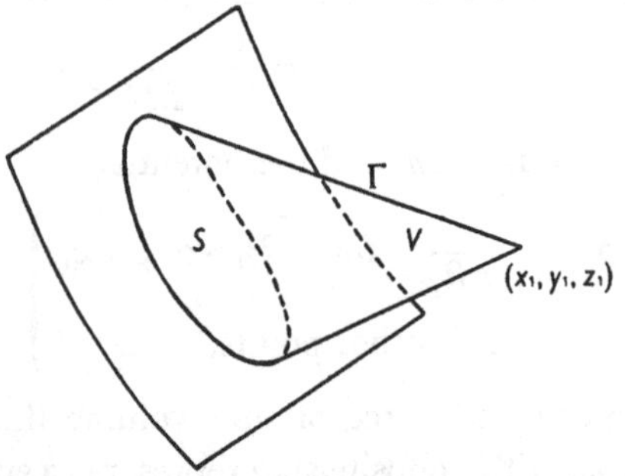

Fig. 3.3. Surface and volume of integration for determining the general solution of the potential equation from the identity (3.6.1).

diverges, but the finite part can be taken, after which the required result follows. However, it is instructive to consider the problem in somewhat greater detail.

Take the surface of integration to consist of

(i) $\Gamma(\epsilon)$, defined by $R_B^2=\epsilon>0$, $z<z_1$,

(ii) $S(\epsilon)$, which is that part of S lying inside $\Gamma(\epsilon)$, and

(iii) S_1, which is that part of the plane $z=z_1-\delta$, $\delta>\epsilon^{\frac{1}{2}}$, lying inside $\Gamma(\epsilon)$. These surfaces are shown diagrammatically in fig. 3.4.

With this surface, bounding the volume $V(\epsilon)$, the identity (3.6.1) gives

$$\int_{S(\epsilon)+\Gamma(\epsilon)+S_1}\phi(\mathbf{R})\frac{B^2(\mathbf{R}-\mathbf{R}_1).\mathbf{n}}{R_B^3}\,dS-\int_{S(\epsilon)+\Gamma(\epsilon)+S_1}\frac{\mathbf{w}(\mathbf{R}).\mathbf{n}}{R_B}\,dS$$
$$=-\int_{V(\epsilon)}\frac{Q(\mathbf{R})}{R_B}\,dV, \tag{3.6.4}$$

for all values of ϵ.

As $\epsilon \to 0$, the first integral over $S(\epsilon)$ is

$$^*\int_S \phi(\mathbf{R}) \frac{B^2(\mathbf{R}-\mathbf{R}_1).\mathbf{n}}{R_B^3} \mathrm{d}S + \frac{A_1}{\epsilon^{\frac{1}{2}}} + o(1), \tag{3.6.5}$$

and the second integral over $S(\epsilon)$, although improper, converges.

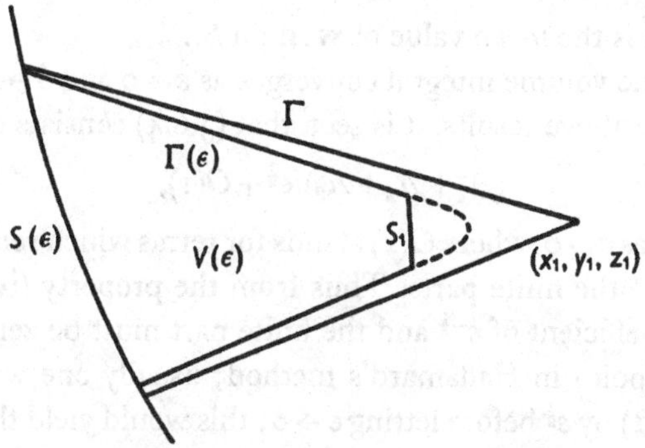

Fig. 3.4. Diagrammatic representation of the surface and volume of integration in the identity (3.6.4).

The direction cosines of the outward normal to $\Gamma(\epsilon)$ are

$$\left.\begin{aligned} &\frac{B^2(x-x_1)}{N}, \quad \frac{B^2(y-y_1)}{N}, \quad -\frac{z-z_1}{N}, \\ \text{where} \quad &N^2 = B^4(x-x_1)^2 + B^4(y-y_1)^2 + (z-z_1)^2 \\ &\quad = (1+B^2)(z-z_1)^2 - B^2\epsilon, \end{aligned}\right\} \tag{3.6.6}$$

and

$$\begin{aligned} \frac{B^2(\mathbf{R}-\mathbf{R}_1).\mathbf{n}}{R_B^3} &= \frac{B^2}{NR_B^3}\{B^2(x-x_1)^2 + B^2(y-y_1)^2 - (z-z_1)^2\} \\ &= -\frac{B^2}{NR_B} = -\frac{B^2}{N\epsilon^{\frac{1}{2}}}. \end{aligned} \tag{3.6.7}$$

Thus, for fixed δ, the contribution from both surface integrals over $\Gamma(\epsilon)$ as $\epsilon \to 0$ is

$$\frac{A_2}{\epsilon^{\frac{1}{2}}} + o(1). \tag{3.6.8}$$

The contribution from S_1 to the first integral in (3.6.4) is

$$\bar{\phi} \int_{S_1} \frac{B^2(\mathbf{R}-\mathbf{R}_1).\mathbf{n}}{R_B^3} \mathrm{d}S = 2\pi\bar{\phi} + \frac{A_3}{\epsilon^{\frac{1}{2}}} + o(1) \tag{3.6.9}$$

from the results of § 3.5, remembering that **n** is outwards from $V(\epsilon)$, and where $\bar{\phi}$ is the mean value of ϕ over S_1 for given δ and $\epsilon = 0$. The second integral over S_1 gives

$$\overline{\mathbf{w}.\mathbf{n}}\int_{S_1}\frac{dS}{R_B} = \overline{\mathbf{w}.\mathbf{n}}\int_0^{\sqrt{(\delta^2-\epsilon)}/B}\int_0^{2\pi}\frac{r}{\sqrt{(\delta^2-B^2r^2)}}\,dr\,d\theta = 2\pi\overline{\mathbf{w}.\mathbf{n}}(\delta-\epsilon^{\frac{1}{2}}), \tag{3.6.10}$$

where $\overline{\mathbf{w}.\mathbf{n}}$ is the mean value of $\mathbf{w}.\mathbf{n}$ on S_1.

Lastly, the volume integral converges as $\epsilon \to 0$ and $\delta \to 0$.

From the above results, it is seen that (3.6.4) consists of terms

$$(A_1 + A_2 + A_3)/\epsilon^{\frac{1}{2}} + O(1), \tag{3.6.11}$$

for fixed δ, as $\epsilon \to 0$, where $O(1)$ stands for terms which remain finite as $\epsilon \to 0$, i.e. the finite parts. Thus from the property (iv) of § 3.4, both the coefficient of $\epsilon^{-\frac{1}{2}}$ and the finite part must be zero. This is the subtle point in Hadamard's method; usually one would multiply (3.6.11) by $\epsilon^{\frac{1}{2}}$ before letting $\epsilon \to 0$; this would yield the relation $A_1 + A_2 + A_3 = 0$, which, though true, is not interesting by itself; its real value is that it ensures the vanishing of the finite part.

Now, in the finite part, if $\delta \to 0$, the mean value of ϕ on S_1 tends uniformly to the value $\phi(\mathbf{R}_1)$, and hence

$$\phi(\mathbf{R}_1) = -\frac{1}{2\pi}\int_V \frac{Q(\mathbf{R})}{R_B}\,dV + \frac{1}{2\pi}\int_S \frac{\mathbf{w}(\mathbf{R}).\mathbf{n}}{R_B}\,dS - \frac{B^2}{2\pi}{}^*\!\!\int_S \phi(\mathbf{R})\frac{(\mathbf{R}-\mathbf{R}_1).\mathbf{n}}{R_B^3}\,dS, \tag{3.6.12}$$

where V and S are in the interior of the dependence domain of $\mathbf{R}_1$, as shown in fig. 3.3 above.

The result (3.6.12) can be written in a different form. If the normal **n** to S has direction cosines l, m, n, then the vector $\boldsymbol{\tau}$,

$$\boldsymbol{\tau} = -\boldsymbol{\Psi}.\mathbf{n} = [-l, -m, B^2 n], \tag{3.6.13}$$

is called the transversal, and

$$\boldsymbol{\tau}.\nabla\phi = -\mathbf{n}.\mathbf{w} = \frac{\partial\phi}{\partial\tau}, \text{ say,} \tag{3.6.14}$$

is called the transversal derivative of ϕ. In terms of transversal derivatives, (3.6.12) is

$$\phi(\mathbf{R}_1) = -\frac{1}{2\pi}\int_V \frac{Q}{R_B}\,dV - \frac{1}{2\pi}{}^*\!\!\int_S \left\{\frac{1}{R_B}\frac{\partial\phi}{\partial\tau} - \phi\frac{\partial}{\partial\tau}\left(\frac{1}{R_B}\right)\right\} dS, \tag{3.6.15}$$

which is Hadamard's form of the result for this special case (Hadamard, 1923).

In order to prove that the equivalent formulae (3.6.12) and (3.6.15) do give the solution of Cauchy's problem, it must be verified that the right-hand sides satisfy the differential equations (3.6.2), or the equivalent equation

$$\frac{\partial^2\phi}{\partial x^2}+\frac{\partial^2\phi}{\partial y^2}-B^2\frac{\partial^2\phi}{\partial z^2}=Q, \qquad (3.6.16)$$

and that ϕ and $\mathbf{w}.\mathbf{n}$ calculated from these formulae reduce to the given values on S. The verification that the differential equation is satisfied is not difficult, but the verification of the boundary values gives some trouble. There are three cases, depending on whether the part of S under consideration is (i) super-inclined, (ii) characteristic, or (iii) sub-inclined. This investigation is not carried out here. It is shown by Hadamard (1923) that the verification is possible in cases (i) and (ii); on a characteristic surface, $\mathbf{w}.\mathbf{n}$ is a tangential component of $\mathbf{v}$, so only ϕ need be given. In case (iii) no verification is possible; the values of ϕ and $\mathbf{w}.\mathbf{n}$ cannot be specified independently on a sub-inclined surface, and the above formulae have to be adapted in some way so that only one of these quantities occurs (for example, by using appropriate Green's functions). A verification may then be possible.

The above solutions are still valid if $\mathbf{v}$ is discontinuous on some characteristic surface or surfaces in V, due to discontinuous boundary data on S, ϕ itself being continuous. For if V is divided into sub-domains by the characteristic surfaces, and (3.6.1) is applied to each sub-domain in turn, then results of the form (3.6.12), but with the left-hand side equal to zero, are obtained for each sub-domain which does not include $\mathbf{R}_1$. The result for the one sub-domain which does include $\mathbf{R}_1$ is exactly of the form (3.6.12). On adding all these formulae, the surface integrals over both sides of the dividing characteristic surfaces cancel, since ϕ and $\mathbf{w}.\mathbf{n}$ must be continuous, and (3.6.12) remains.

3·7. The general solution of the vector equations

Just as for subsonic flow, the first-order vector equations

$$\nabla \wedge \mathbf{v} = \boldsymbol{\zeta}, \quad \nabla . \mathbf{w} = Q, \quad \mathbf{w} = \boldsymbol{\Psi} . \mathbf{v}, \tag{3.7.1}$$

for supersonic flow, can be integrated directly† by using the vector identity (see Appendix 1)

$$\int_S (\mathbf{v}_1 \mathbf{w}_2 . \mathbf{n} + \mathbf{v}_2 \mathbf{w}_1 . \mathbf{n} - \mathbf{v}_1 . \mathbf{w}_2 \mathbf{n}) \, dS$$
$$= \int_V \{\mathbf{v}_1 \nabla . \mathbf{w}_2 + \mathbf{v}_2 \nabla . \mathbf{w}_1 - \mathbf{w}_1 \wedge (\nabla \wedge \mathbf{v}_2) - \mathbf{w}_2 \wedge (\nabla \wedge \mathbf{v}_1)\} \, dV. \tag{3.7.2}$$

Take $\mathbf{v}_1$, $\mathbf{w}_1$ to be the required solutions of (3.7.1) at the point $\mathbf{R}_1$, take

$$\left.\begin{aligned} \mathbf{v}_2 &= \nabla\left(\frac{1}{R_B}\right) \quad \text{for} \quad R_B^2 > 0, \ z < z_1, \\ &= 0 \ \text{at all other points}, \end{aligned}\right\} \tag{3.7.3}$$

and apply the identity (3.7.2) to the volume, V, enclosed by the cone, Γ ($R_B = 0, z < z_1$), a surface S upstream from $\mathbf{R}_1$, and a spherical surface, S_1, of small radius with centre at $\mathbf{R}_1$. The resulting integrals diverge, but the integrands satisfy the conditions of § 3.4, so their finite parts can be taken.

The surface integral over S_1 is

$$\bar{\mathbf{v}} \, {}^*\!\!\int_{S_1} \mathbf{w}_2 . \mathbf{n} \, dS + \bar{\mathbf{w}} \wedge {}^*\!\!\int_{S_1} \mathbf{v}_2 \wedge \mathbf{n} \, dS, \tag{3.7.4}$$

where the bars denote mean values over S_1. The results of § 3.5 show that the value of the first term in (3.7.4) is $2\pi\bar{\mathbf{v}}$. The integral in the second term of (3.7.4) is (cf. (2.7.4) and (2.7.5))

$${}^*\!\!\int_{S_1} \frac{M^2(z - z_1)}{R_B^3} \mathbf{k} \wedge \mathbf{n} \, dS, \tag{3.7.5}$$

and on transforming to polar angles ϑ, φ, with polar axis parallel to the z-axis, this becomes

$${}^*\!\!\int_0^{\cos^{-1}(B/M)} \int_0^{2\pi} \frac{M^2 \cos\vartheta (\sin\vartheta \sin\varphi \mathbf{i} - \sin\vartheta \cos\varphi \mathbf{j}) \sin\vartheta}{(M^2 \cos^2\vartheta - B^2)^{\frac{3}{2}}} \, d\vartheta \, d\varphi = 0. \tag{3.7.6}$$

The finite part of the integral over Γ vanishes, and the integral over S can be written

$${}^*\!\!\int_S \{(\mathbf{n} . \mathbf{w}) \mathbf{v}_2 + (\mathbf{n} \wedge \mathbf{v}) \wedge \mathbf{w}_2\} \, dS. \tag{3.7.7}$$

† Ward (1952 *a*, *b*).

Hence, by collecting all these results, and letting the radius of S_1 tend to zero, the finite part of (3.7.2) for the specified domains of integration gives

$$\mathbf{v}(\mathbf{R}_1) = -\frac{1}{2\pi}{}^*\int_S \mathbf{n}.\mathbf{w}(\mathbf{R})\nabla\left(\frac{1}{R_B}\right)\mathrm{d}S - \frac{B^2}{2\pi}{}^*\int_S \frac{\{\mathbf{n}\wedge\mathbf{v}(\mathbf{R})\}\wedge(\mathbf{R}-\mathbf{R}_1)}{R_B^3}\mathrm{d}S$$
$$+\frac{1}{2\pi}{}^*\int_V Q(\mathbf{R})\nabla\left(\frac{1}{R_B}\right)\mathrm{d}V - \frac{B^2}{2\pi}{}^*\int_V \frac{(\mathbf{R}-\mathbf{R}_1)\wedge\boldsymbol{\zeta}(\mathbf{R})}{R_B^3}\mathrm{d}V, \quad (3.7.8)$$

where S and V are in the interior of the dependence domain of $\mathbf{R}_1$, as shown in fig. 3.3 of §3.6.

To prove that (3.7.8) is the solution of (3.7.1) for the given boundary values on S, it must be verified that the right-hand side satisfies these equations, and that $\mathbf{n}.\mathbf{w}$ and $\mathbf{n}\wedge\mathbf{v}$ reduce to the specified values on S. These verifications are subject to the same remarks made at the end of §3.6. If S is super-inclined or characteristic everywhere, then (3.7.8) gives the complete solution for the properly set Cauchy problem; but if part of S is sub-inclined, then $\mathbf{n}.\mathbf{w}$ and $\mathbf{n}\wedge\mathbf{v}$ cannot be specified independently there.

3.8. Surface and line distributions of vorticity

The velocity field

$$\mathbf{v}(\mathbf{R}_1) = -\frac{B^2}{2\pi}{}^*\int_V \frac{(\mathbf{R}-\mathbf{R}_1)\wedge\boldsymbol{\zeta}(\mathbf{R})}{R_B^3}\mathrm{d}V, \qquad (3.8.1)$$

where V is the whole of the dependence domain of $\mathbf{R}_1$, satisfies the vector equations of motion (3.7.1) in the case when $Q=0$ (that is, when there are no sources) provided that the finite part of the integral converges in the ordinary sense. This formula was given by Robinson (1948*b*), who obtained it by a different method.

The investigation of surface and line distributions of vorticity in supersonic flow follows the same lines as the corresponding investigation for subsonic flow in §2.8. If $\boldsymbol{\omega}$ is the surface density of vorticity on a vortex sheet, S, then

$$\mathbf{v}(\mathbf{R}_1) = -\frac{B^2}{2\pi}{}^*\int_S \frac{(\mathbf{R}-\mathbf{R}_1)\wedge\boldsymbol{\omega}(\mathbf{R})}{R_B^3}\mathrm{d}S, \qquad (3.8.2)$$

the integration being over that part of S which lies in the dependence domain of $\mathbf{R}_1$. The velocity is discontinuous at S; if $\mathbf{v}_1$ and $\mathbf{v}_2$ are the velocities on either side of S at the same point, and $\mathbf{n}$ is the normal to S at the point, directed from side 1 to side 2, then it can be shown from (3.8.2), or directly from (3.7.1), that

$$\mathbf{v}_1-\mathbf{v}_2=\mathbf{n}\wedge\boldsymbol{\omega} \tag{3.8.3}$$

and

$$\boldsymbol{\omega}=\mathbf{n}\wedge(\mathbf{v}_2-\mathbf{v}_1), \tag{3.8.4}$$

which are identical with the results for subsonic flow (cf. (2.8.5) and (2.8.6)).

For a line vortex, L, of strength K,

$$\mathbf{v}(\mathbf{R}_1)=-\frac{B^2K^*}{2\pi}\int_L \frac{(\mathbf{R}-\mathbf{R}_1)\wedge d\mathbf{s}}{R_B^3}, \tag{3.8.5}$$

the integration being along those parts of L which lie in the dependence domain of $\mathbf{R}_1$.

3.9. Horseshoe vortices

The field of a horseshoe vortex, consisting of a bound vortex with positive circulation K lying along the x-axis for $-a \leqslant x \leqslant a$, and two semi-infinite free vortices trailing downstream parallel to the z-axis from the ends of the bound vortex, can be calculated from (3.8.5). The velocity vanishes for $z<0$, while for $z>0$, the rectangular components of $\mathbf{v}(\mathbf{R})$ are

$$u=\frac{K}{2\pi}\left\{\frac{yz}{[(x+a)^2+y^2]\sqrt{[z^2-B^2(x+a)^2-B^2y^2]}} - \frac{yz}{[(x-a)^2+y^2]\sqrt{[z^2-B^2(x-a)^2-B^2y^2]}}\right\}, \tag{3.9.1}$$

$$v=\frac{K}{2\pi}\left\{\frac{(x-a)z}{[(x-a)^2+y^2]\sqrt{[z^2-B^2(x-a)^2-B^2y^2]}} - \frac{(x+a)z}{[(x+a)^2+y^2]\sqrt{[z^2-B^2(x+a)^2-B^2y^2]}} - \frac{B^2z}{z^2-B^2y^2}\times\left[\frac{x-a}{\sqrt{[z^2-B^2(x-a)^2-B^2y^2]}} - \frac{x+a}{\sqrt{[z^2-B^2(x+a)^2-B^2y^2]}}\right]\right\}, \tag{3.9.2}$$

$$w=\frac{KB^2}{2\pi}\frac{y}{z^2-B^2y^2}\left\{\frac{x-a}{\sqrt{[z^2-B^2(x-a)^2-B^2y^2]}} - \frac{x+a}{\sqrt{[z^2-B^2(x+a)^2-B^2y^2]}}\right\}, \tag{3.9.3}$$

where all terms are present when real, and are to be omitted wherever they become imaginary. The above results show that the effects of such a horseshoe vortex are confined to the influence domains of the *tips* of the bound vortex, and thus there is no field from the bound vortex itself outside these domains, which is a rather surprising result. The reason for this is discovered when a superposition of such horseshoe vortices in the z-direction is considered. It is then found that the disturbance is non-zero in the region bounded by the characteristic planes which contain the foremost and hindmost bound vortices and by the characteristic cones from the tips of the foremost bound vortex. The cross-section of this domain by a plane $z = constant$ is shown shaded in fig. 3.5. As the width of the distribution in the z-direction becomes very small, the

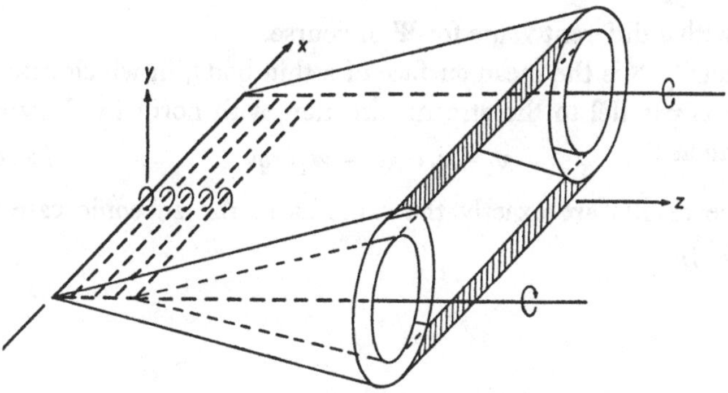

Fig. 3.5. Characteristic surfaces associated with a superposition of simple horseshoe vortices.

disturbance is practically confined to the upstream boundary of the shaded domain in fig. 3.5, and disappears in the limit of zero width.

More elaborate distributions of horseshoe vortices can be used to represent the flow over supersonic wings just as in subsonic theory. This method has been used by Schlichting (1936) to calculate the supersonic flow past the tips of wings of rectangular planform, and by Robinson and Hunter-Tod (1947*a*) to calculate the down-wash in the wake of a delta wing.

3.10. Surface distributions of sources

It follows from (3.6.12) and (3.7.8), respectively, that the potential and the velocity field for a surface distribution of sources on a surface, S, with surface density, q, are

$$\phi(\mathbf{R}_1) = -\frac{1}{2\pi}\int_S \frac{q(\mathbf{R})}{R_B}\,dS, \tag{3.10.1}$$

and
$$\mathbf{v}(\mathbf{R}_1) = \frac{1}{2\pi} \overset{*}{\int_S} q(\mathbf{R}) \nabla\left(\frac{1}{R_B}\right) dS, \tag{3.10.2}$$

where the integrations are over those parts of S which lie in the dependence domain of the point $\mathbf{R}_1$.

The potential is continuous and the normal component of $\mathbf{v}$ is discontinuous on S, the expressions for the discontinuities of $\mathbf{v}$ and $\mathbf{w}$ being formally the same as in the subsonic case (§ 2.10), namely,
$$\mathbf{v}_2 - \mathbf{v}_1 = \frac{q\mathbf{n}}{\mathbf{n}.\mathbf{\Psi}.\mathbf{n}}, \quad \mathbf{w}_2 - \mathbf{w}_1 = \frac{q\mathbf{\Psi}.\mathbf{n}}{\mathbf{n}.\mathbf{\Psi}.\mathbf{n}}, \tag{3.10.3}$$

but with a different value for $\mathbf{\Psi}$ of course.

Usually S is the mean surface of a thin body, in which case the layer is parallel to the stream direction, with normal $\boldsymbol{\nu}$ ($\mathbf{U}.\boldsymbol{\nu} = 0$), and then
$$\mathbf{v}_2 - \mathbf{v}_1 = \mathbf{w}_2 - \mathbf{w}_1 = q\boldsymbol{\nu}. \tag{3.10.4}$$

These results are exactly the same as in the subsonic case (cf. (2.10.8)).

CHAPTER 4

BOUNDARY CONDITIONS, AERODYNAMIC FORCES, UNIQUENESS AND FLOW-REVERSAL THEOREMS

4.1. Boundary conditions at solid and free surfaces

Provided that the dimensions of a solid body immersed in a real gas are all very large compared with the mean free path of the molecules in its vicinity, the boundary condition at the surface of the body can be taken to be that the relative velocity of the continuous fluid is zero at the surface; that is, that the fluid adheres to the surface.† The theory and experimental observation of the flow near a solid surface shows that in general the effects of viscosity and heat conduction are effectively confined to a thin boundary layer, provided that the viscosity and heat conductivity are sufficiently small. The smaller the viscosity, the thinner is the boundary layer, and in the limit as viscosity and heat conductivity tend to zero, this layer becomes a vortex sheet coincident with the body surface. In this case, which is the state of affairs considered here, nothing can be said about the component of velocity tangential to the surface, since the vorticity density in the vortex sheet is not known *a priori*. Thus the boundary condition for inviscid flow is that the normal component of the relative fluid velocity at the surface is zero, and slip may occur. This assumes that no separation of the boundary layer occurs at any point of the surface, which is an assumption that sometimes requires a more detailed formulation, as at sharp trailing edges for example (see §4.2). If $\mathbf{n}$ is the unit normal to the body surface, the mathematical expression of this condition is

$$\mathbf{n}.(\mathbf{U}+\mathbf{v})=0. \tag{4.1.1}$$

It has been shown in §1.9 that this last condition can be replaced in linearized theory by

$$\boldsymbol{\nu}.\mathbf{v}=-\mathbf{n}.\mathbf{U}, \tag{4.1.2}$$

where, for thin bodies, $\boldsymbol{\nu}$ is the unit normal to the mean body surface, and (4.1.2) is to be satisfied on the mean surface.

† For a discussion of this condition and the evidence that supports it, see *Modern Developments in Fluid Dynamics*, ed. S. Goldstein (Oxford, 1938), vol. 2, pp. 676–80.

At free boundaries, where the fluid is in contact with another fluid or the same fluid in a different state of motion, any relative velocity causes a shear layer to develop between the two fluids. This shear layer may be laminar, in which case it is controlled by the viscosity of the fluids, or it may be turbulent, in which case the fluids mix and the viscosity plays a minor role. In the usual idealized theory of such free boundaries between two inviscid fluids, it is assumed that the shear layer is replaced by a vortex sheet and that mixing does not occur. Provided that the dimensions of the free surface in the stream direction are not too great, the shear layer is laminar, and this assumption gives a correct account of the state of affairs in the limit as viscosity tends to zero. But when the free surface extends over great distances in the stream direction, the inherent instability of the vortex sheet causes a turbulent shear layer to develop at some finite distance from the upstream boundary of the vortex sheet, the actual distance depending largely on the initial unsteadiness or turbulence of the streams, and their relative velocity. It may be supposed that for perfectly steady initial flows no turbulent shear layer would occur, but this condition is impossible in practice, and for any non-zero viscosity and initial turbulence, however small, there is always a place where the shear layer becomes important, so the limiting (vortex sheet) form of the free surface is not attained uniformly at great distances downstream. Thus any solution obtained on the vortex sheet hypothesis cannot represent a physically real flow, but in spite of this such solutions can be useful, and the assumption is adopted here.

The boundary conditions at the free surface are that the normal velocity must be zero on both sides of it in steady flow, and that the pressure must be continuous across it.

If the fluid is at rest on one side of the free boundary, which is the usual case, and the only one considered here, then the pressure at the free boundary must be constant. Also, if $\mathbf{n}$ is the unit normal to the boundary, then

$$\mathbf{n}.(\mathbf{U}+\mathbf{v})=0. \tag{4.1.3}$$

In linearized theory, $\mathbf{v}$ is small, and (4.1.3) then shows that $\mathbf{U}.\mathbf{n}$ must be small, so the free boundary must be nearly parallel to $\mathbf{U}$ at every point. Just as for solid boundaries, there is a mean free surface whose generators are parallel to $\mathbf{U}$, such that the actual free surface lies at a small distance from it at a finite distance downstream from its commencement. The boundary condition of constant pressure can be applied on this surface without causing any greater errors than are inherent in the assumption of no shear layer and in the linearized approximation, and the linearized Bernoulli's equation can be used for the pressure. Finally, then, the two boundary conditions reduce to the combined condition

$$\mathbf{U}.\mathbf{v} = \text{known constant} \tag{4.1.4}$$

on the mean free boundary. If the free boundary extends upstream to infinity, then this constant is zero.

4.2. Vortex wakes and the Kutta-Joukowski condition

For bodies with a sharp trailing edge, experimental observation of subsonic flows shows that, in general, the action of viscosity causes the flow to leave the trailing edge smoothly, and that a thin wake is formed downstream from the trailing edge by the retarded layers of fluid from the body surface. As the viscosity tends to zero, this situation is idealized by the assumptions that the wake is infinitely thin as it leaves the body, and that the velocity at the sharp trailing edge is finite. This last condition is due to Kutta and Joukowski and bears their names. It is assumed here that this Kutta-Joukowski condition applies not only at sharp trailing edges in subsonic flows, but also at sharp subsonic trailing edges in supersonic flows; the condition is found to be satisfied automatically at supersonic trailing edges.

The velocity varies rapidly through the thin wake in real flows, so the idealized wake can be taken as a vortex sheet from the time it leaves the trailing edge. At a distance from the body this vortex sheet is supposed to roll up under the action of its own induced velocity, until it ultimately assumes the form of distinct vortex cores of finite diameter at large distances downstream. The greater the strength of the vortex sheet the more rapidly does it roll up. Experimental evidence generally supports this supposition, but the action of viscosity ultimately dissipates the vortex cores, so they do not extend downstream to 'infinity' in practice.

Linearized theory applies to bodies whose incidence, camber, thickness ratio, etc., are all small, so all edges of thin bodies ultimately become sharp. It is found later that at subsonic edges the perturbation velocity becomes infinite in general; series expansions like (1.8.4) are non-uniformly convergent, and the assumption of smallness for the perturbation velocity is violated in neighbourhoods of these edges. However, the flow outside these neighbourhoods can be shown to give the correct approximation to the exact inviscid flow in many cases,† and it appears to be a reasonable assumption that this is true always. This assumption has important consequences in the calculation of aerodynamic forces (see §§4.6 and 4.7).

In linearized theory, the Kutta-Joukowski condition is that the component of the perturbation velocity normal to the mean body surface must be finite in a neighbourhood of any trailing edge.‡ This causes the appearance in the mathematical solution of surfaces stretching downstream from the trailing edges on which a tangential component of velocity is discontinuous. These are the vortex wakes described above, and certain physical conditions must be satisfied on them. These conditions are that the velocity must be purely tangential on either side, and that the pressure must be continuous there. If $\mathbf{v}_1$, $\mathbf{v}_2$ are the perturbation velocities on either side of the vortex sheet and $\mathbf{n}$ is the normal to the sheet, then the above conditions are (cf. §1.7)

$$\mathbf{n}.(\mathbf{U}+\mathbf{v}_1)=\mathbf{n}.(\mathbf{U}+\mathbf{v}_2)=0, \tag{4.2.1}$$

and, if the perturbation velocities are small,

$$\mathbf{U}.\mathbf{v}_1+\tfrac{1}{2}\mathbf{v}_1.\mathbf{w}_1=\mathbf{U}.\mathbf{v}_2+\tfrac{1}{2}\mathbf{v}_2.\mathbf{w}_2 \tag{4.2.2}$$

approximately, from the quadratic approximation to Bernoulli's equation. This last equation can be written

$$(\mathbf{v}_1-\mathbf{v}_2).(\mathbf{U}+\tfrac{1}{2}\mathbf{w}_1+\tfrac{1}{2}\mathbf{w}_2)=0, \tag{4.2.3}$$

or, if $\boldsymbol{\omega}$ is the surface vorticity, from (2.8.5) and (3.8.3),

$$\mathbf{n}\wedge\boldsymbol{\omega}.(\mathbf{U}+\tfrac{1}{2}\mathbf{w}_1+\tfrac{1}{2}\mathbf{w}_2)=0. \tag{4.2.4}$$

Thus, either $\boldsymbol{\omega}=0$ and there is no vortex sheet at all, or $\mathbf{n}$, $\boldsymbol{\omega}$ and $\mathbf{U}$ are approximately co-planar. Now $\mathbf{n}$ and $\boldsymbol{\omega}$ must be orthogonal vectors, and from (4.2.1) it is seen that $\mathbf{n}$ and $\mathbf{U}$ are approximately

† R. T. Jones (1950), M. J. Lighthill (1951).

‡ The author is indebted to Professor S. Goldstein for pointing out that the usual condition of finite velocity is too restrictive in linearized theory.

orthogonal vectors, so $\boldsymbol{\omega}$ must be approximately parallel to $\mathbf{U}$. It is usual to adopt Prandtl's assumption, that the vorticity $\boldsymbol{\omega}$ is exactly parallel to $\mathbf{U}$ at all points on the vortex sheet. Then the vortex lines are straight, and the vortex sheet is a surface with generators and edges parallel to $\mathbf{U}$, which extends downstream to infinity from the trailing edge. This surface is the linearized vortex sheet.

That the parallelism of $\boldsymbol{\omega}$ and $\mathbf{U}$ is only approximate on the above arguments is a consequence of using the quadratic approximation to Bernoulli's equation in (4.2.2); if the linear approximation is used, then $\boldsymbol{\omega}$ and $\mathbf{U}$ are exactly parallel, in accordance with the result obtained in § 1.12 (where it has been shown that the component of free vorticity in the direction of $\mathbf{U}$ must be constant, and hence that free vortex lines and free vortex sheets must be parallel to $\mathbf{U}$). Thus Prandtl's assumption is consistent with the results of linearized theory as developed above. The use of the quadratic approximation for the pressure is not consistent with this assumption in general.

If $\boldsymbol{\nu}$ is the unit normal to the linearized vortex sheet, then the boundary conditions on it are

$$\boldsymbol{\nu}.(\mathbf{v}_1-\mathbf{v}_2)=0, \tag{4.2.5}$$

$$\mathbf{U}.(\mathbf{v}_1-\mathbf{v}_2)=0. \tag{4.2.6}$$

With these boundary conditions, it is found that $\boldsymbol{\omega}$ and $\mathbf{v}$ become infinite in general at the edges (if any) of the vortex sheet, so (4.2.2) and (4.2.6) are not valid near to and at the edges, and the above assumptions become difficult to justify. In addition, it is found later that these infinite velocities at the edges give rise to forces which tend to 'tear the vortex sheet apart', and which do not arise in practice. Thus the linearized vortex sheet cannot possibly represent a true approximation to the real state of affairs in a wake, and the discrepancies merit further discussion.

It appears from (4.2.1) that the wake cannot extend to infinity as assumed, since the streamlines do not lie in the linearized vortex sheet, even in linearized theory. As mentioned above, the vortex sheet rolls up to form vortex cores, and the rate of rolling up depends upon the vortex strength. At a fixed distance behind the body, the vortex wake will ultimately have the form of the assumed linearized vortex sheet as the strength tends to zero, while, on the other hand, for a given strength, however small, a distance can always be found at which the vortex sheet has rolled up to a given extent. On the present assumptions, the linearized vortex sheet represents an approximation valid in a region which increases as the strength decreases, but not valid uniformly in the whole of space. The physical meaning of the infinite vortex strength at edges is not at all clear. It may be that the maximum vortex strength tends to zero more slowly than a mean vortex strength, and that it is located nearer and nearer to the edge as this mean vortex strength tends to zero. The removal of this difficulty requires further investigation, both for its own sake and because the occurrence of the infinite vortex strength causes a further difficulty in the calculation of aerodynamic forces (see §4.6).

4.3. Shocks and expansion waves

In actual supersonic flows, the wave fronts of elementary compression waves tend to pile up on one another to form very sharp wave fronts, or shocks, at which the velocity decreases and the density and pressure increase very rapidly as the fluid passes through them. Due to the very large velocity gradients in such shocks, there is a large viscous dissipation of energy and consequent heat conduction inside them. This energy dissipation is non-reversible, and the entropy of the fluid increases as it passes through the shock. In the limit, as viscosity and heat conduction tend to zero, these shocks become infinitely thin, and the velocity is discontinuous at them. The energy dissipation is finite in the limit, so there is also a finite entropy change. The equations of motion for an inviscid fluid are not satisfied at such discontinuities in the flow, and the equations of conservation of mass, momentum and energy have to be derived as difference equations across the shock, as in §1·7. The solutions of these difference equations admit expansion shocks, in addition to compression shocks, but a calculation of the entropy change shows that the entropy decreases through the hypothetical expansion shocks, which is contrary to the Second Law of Thermodynamics. Thus expansion shocks cannot have any physical reality.

The expansion waves that occur in actual flows extend over finite regions as a consequence of the fact that the wave fronts of elementary expansion waves tend to separate. Thus there is no rapid change in velocity through such expansion waves; viscosity and heat conduction have little effect on their structure, and the flow through them is isentropic in the limit as viscosity and heat conductivity tend to zero.

For the exact theory of shocks and expansion waves in an inviscid fluid, the reader is referred to the book *Supersonic Flow and Shock Waves*, by R. Courant and K. O. Friedrichs (Interscience Publishers, New York, 1948), where a very detailed account of the phenomena may be found.

As shown in § 1.11, discontinuities of the velocity and pressure occur on characteristic surfaces of the equations of motion in linearized theory, and compressive discontinuities of this kind usually represent shocks. The exact equations for a shock discontinuity show that, as the magnitude of the discontinuity tends to zero, the position of the shock approaches a characteristic surface. Two difficulties remain however: first, this approach is not uniform at great distances from the disturbance which causes the shock, and secondly, the change of entropy at the shock appears to contradict the assumption of isentropic flow along streamlines, on which linearized theory is based. The first difficulty is merely another manifestation of the circumstance noticed already, that linearized theory does not give a true picture of the flow at great distances from bodies. The second difficulty requires further consideration because it could affect the flow near bodies.

If p_1 and p_2 are respectively the pressures just upstream and just downstream from a point of the shock, then the strength of the shock is usually defined as $(p_2 - p_1)/p_1$. If the shock strength is small, it is of the same order as the velocity change divided by the upstream velocity, and it can be shown that the entropy increase is proportional to the cube of the strength. Thus to a linear approximation (and, indeed, to a second approximation in which squares of small quantities are retained in the equations of motion) the flow through a shock is isentropic along its streamlines.

The discontinuities on characteristics in linearized theory can be expansive as well as compressive, and it can be shown that centred continuous expansion waves tend to become discontinuous as their strengths tend to zero, but not uniformly at great distances from the centre. The entropy decrease through the hypothetical expansion shocks mentioned above is proportional to the cube of the strength, when small, and the flow is again isentropic in linearized theory. Thus conditions near the body are not affected by replacing the continuous expansion waves by weak discontinuous expansions.

If $\mathbf{n}$ is the unit normal to a shock surface in linearized theory, then the conditions at the surface are (see § 1.11) that the normal component of $\mathbf{w}$ and the tangential component of $\mathbf{v}$ must be continuous, and that Bernoulli's equation has the same form on both sides of the surface; that is,

$$\mathbf{n}.\mathbf{w} \text{ is continuous,} \tag{4.3.1}$$

$$\mathbf{n}\wedge\mathbf{v} \text{ is continuous} \tag{4.3.2}$$

and
$$p-p_0 = -\rho_0\mathbf{U}.\mathbf{v} \text{ on both sides of the shock.} \tag{4.3.3}$$

The condition (4.3.2) is part of the momentum condition across the discontinuity, and, from (1.7.2), the linearized form of the second part of this condition is that

$$p-p_0+(\rho-\rho_0)(\mathbf{n}.\mathbf{U})^2+\rho_0\mathbf{n}.\mathbf{U}(\mathbf{U}+\mathbf{v}).\mathbf{n} \tag{4.3.4}$$

must be continuous. On substituting the values of $p-p_0$ from (4.3.3) and $\rho-\rho_0$ from (1.11.20), this last condition is that

$$\rho_0(\mathbf{n}.\mathbf{U})(\mathbf{n}.\mathbf{w})+\rho_0(\mathbf{n}\wedge\mathbf{U}).(\mathbf{v}\wedge\mathbf{n}) \tag{4.3.5}$$

must be continuous, which is satisfied automatically by virtue of (4.3.1) and (4.3.2). Thus this fourth condition adds nothing new in linearized theory, having been replaced effectively by the condition of isentropic flow through the shock.

In the case of extremely weak shocks, the velocity as given by linearized theory may be continuous, and the shocks manifest themselves in the form of infinite space derivatives of the velocity components at the appropriate characteristic surfaces.

Finally, at all points in the fluid at which there are no shocks and no vortex sheets, the velocity, $\mathbf{v}$, must be continuous. Also the potential ϕ must be continuous at shocks, since (4.3.2) allows only a constant discontinuity and this can be made zero without loss of generality. Thus ϕ can be taken to be continuous at all points in the fluid at which there are no vortex sheets, and is conveniently taken to be zero at infinity upstream from bodies.

4.4. Conditions at great distances from bodies

The arguments presented above show that linearized theory in its present form cannot give accurate information about the flow at great distances from bodies. From the mathematical point of view, this non-uniformity of the approximation is due to the non-uniform convergence at infinity, and at characteristic surfaces of discontinuity, of the series for Φ in powers of t. The work of Lighthill (1949*b*) and Whitham (1950, 1952) shows that modified forms of linearized theory (based on expansions in powers of t (and $\log t$) which are different from those of § 1.8) can be constructed which give a uniform first approximation to the flow conditions in regions where the present theory fails. This work also shows that the present theory does give the correct first approximation on and near bodies (except near sharp leading edges), so the incorrect solutions at infinity do not have a disastrous effect on the whole solution. In spite of the deficiencies of the linearized solutions, a knowledge of their general behaviour at great distances from bodies is useful, particularly for the proof of uniqueness. This behaviour is different for subsonic and supersonic flows.

The solutions of the linearized equations of motion for boundary conditions of the form derived above are not unique without a further assumption. The simplest form of this assumption is that the aerodynamic forces on any finite area of a body surface is finite; it follows from this that the overall aerodynamic force on a body of finite dimensions is finite. The reasonableness of this assumption seems to require little justification. A consequence of the assumption, which is equivalent to it, is that the perturbation velocity potential, ϕ, is finite at all points on the surface of a finite body, and at all points in the surrounding fluid. If r is distance measured normally from an edge of the body, then it is shown in § 4.6 that a further consequence of the condition is that $|r^{\frac{1}{2}}\mathbf{v}|$ must be bounded as $r \to 0$ at all points on the edge.

For subsonic flow, if there are no vortex sheets, then the velocity field for a finite body can be represented by a suitable distribution of sources, the algebraic sum of the source strengths being zero if the body is closed. The velocity field at great distances can be represented as a series of terms representing a source field, a doublet field, a quadrupole field, etc.; since the source term vanishes, the dominant term is the doublet field. If R is the distance from some fixed origin in or near the body, then $|R^2\phi|$ and $|R^3\mathbf{v}|$ are bounded as $R\rightarrow\infty$. On the other hand, if there are vortex sheets downstream from the body, then Kelvin's circulation theorem shows that the total vortex strength in the sheets must be zero. Thus at large distances from the body and the vortex sheets, the velocity field must approximate to that of a semi-infinite vortex pair, and since the overall aerodynamic force on the body is finite, this vortex pair must be of finite strength. If r is the distance from the nearest point of the body and the vortex sheet, then the expressions in § 2.9 for the velocity field of a horseshoe vortex show that $|r^2\mathbf{v}|$ is bounded as $r\rightarrow\infty$, and that $|R^2\mathbf{U}.\mathbf{v}|$ is bounded as $R\rightarrow\infty$.

For supersonic flow, disturbances cannot be propagated upstream, so the perturbation velocity must vanish upstream from the characteristic surface which bounds the influence domain of the body. This characteristic surface is the envelope of downstream characteristic half-cones with vertices on the leading edges of the body.

4.5. Summary of the boundary conditions

For convenience, the boundary conditions derived in §§ 4.1–4.4 are collected together in summary form below.

(i) If $\mathbf{n}$ is the normal to the body surface and $\boldsymbol{\nu}$ is the normal to the mean body surface, then

$$\boldsymbol{\nu}.\mathbf{v}=-\mathbf{n}.\mathbf{U}$$

on the mean body surface for thin bodies.

(ii) On the mean surface corresponding to a free boundary, outside which the fluid is at rest with known pressure p_1,

$$\mathbf{U}.\mathbf{v}=(p_0-p_1)/\rho_0.$$

(iii) At sharp trailing edges, the Kutta-Joukowski condition, that $\boldsymbol{\nu}.\mathbf{v}$ is bounded, must be applied.

(iv) In the wake of the body, any vortex sheets extending downstream from sharp trailing edges are cylindrical surfaces whose generators and edges are parallel to $\mathbf{U}$. These surfaces are the linearized vortex sheets, and if $\boldsymbol{\nu}$ is the normal at any point ($\boldsymbol{\nu}.\mathbf{U}=0$), then $\boldsymbol{\nu}.\mathbf{v}$ and $\mathbf{U}.\mathbf{v}$ are continuous through the surfaces.

(v) In supersonic flow, on characteristic surfaces at which $\mathbf{v}$ is discontinuous, ϕ, $\mathbf{n}.\mathbf{w}$ and $\mathbf{n}\wedge\mathbf{v}$ are continuous.

(vi) At all other points in the fluid ϕ and $\mathbf{v}$ are continuous.

(vii) The aerodynamic force on any finite area of the body surface is finite, and the overall aerodynamic force on a body of finite dimensions is finite; or, ϕ is finite at all points on the surface of a finite body, and if r is distance measured normally from an edge of the body, then $|r^{\frac{1}{2}}\mathbf{v}|$ is bounded as $r\to 0$.

(viii) In subsonic flow, if R is the distance from a fixed origin in or near the body, then, when there are no vortex sheets, $|R^3\mathbf{v}|$ is bounded as $R\to\infty$. When there are vortex sheets, if r is the distance from the nearest point of the body and the vortex sheets, then $|r^2\mathbf{v}|$ is bounded as $r\to\infty$, and $|R^2\mathbf{U}.\mathbf{v}|$ is bounded as $R\to\infty$.

(ix) In supersonic flow, $\mathbf{v}=0$ at all points upstream from the characteristic surface which bounds the influence domain of the body.

4.6. Aerodynamic forces

The calculation of the overall aerodynamic force on a body can be carried out in two ways, either by integrating the pressure over the body surface, or by integrating the momentum flux across some suitably chosen surface which encloses the body. These two methods are theoretically equivalent, but, in linearized theory, the occurrence of infinite velocities in various places causes some difficulties. These infinite velocities arise as a consequence of applying the linearized boundary conditions, as explained above, and cause trouble in both methods. If the direct method of integrating the normal pressures is adopted, then, first, the usual approximations (either linear or quadratic) to Bernoulli's equation are certainly not valid in neighbourhoods of the singularities. Secondly, in cases where more exact solutions are known, it is found that the process of letting the body thickness tend to zero leaves forces on sharp edges in the limit, which linearized theory does not give.

These difficulties do not arise when the momentum-flux method is used. The momentum flux can be evaluated at a surface which lies at some distance from the edges at which infinite velocities occur, so that the approximate expressions for the pressure are valid. It is then found that the forces on sharp edges are given correctly. However, another difficulty arises in this method from the infinite velocity at edges of vortex sheets; this is dealt with below.

For steady inviscid flow, with no impressed field of force, Euler's momentum theorem (1.2.7) for any closed surface, S, inside which there are no sources and $\mathbf{u}$ is not singular, gives

$$\int_S (p\mathbf{n}+\rho\mathbf{u}\mathbf{u}.\mathbf{n})\,\mathrm{d}S=0, \tag{4.6.1}$$

where $\mathbf{n}$ is taken to be the outward normal to S.

Now, without further approximation, let S consist of the body surface, Σ^*, any surface, S', which completely encloses the body, and a surface, σ^*, which lies close to both sides of any vortex sheets in the fluid between Σ^* and S'. Since $\mathbf{u}.\mathbf{n}=0$ on the body surface, Σ^*, the contribution to (4.6.1) from Σ^* is

$$\mathbf{F}=\int_{\Sigma^*} p\mathbf{n}\,\mathrm{d}S, \tag{4.6.2}$$

where $\mathbf{F}$ is the aerodynamic force on the body. The contribution from σ^* vanishes, since $\mathbf{u}.\mathbf{n}=0$, and p is continuous at a vortex sheet. Hence

$$\mathbf{F}=-\int_{S'} (p\mathbf{n}+\rho\mathbf{u}\mathbf{u}.\mathbf{n})\,\mathrm{d}S. \tag{4.6.3}$$

This integral is invariant for changes of S' such that S' always encloses the body, as is seen from its method of derivation.

By letting the thickness, etc., of the body become very small, if a surface, S', exists on which the perturbation velocity, $\mathbf{v}$, is uniformly small compared with the stream velocity, $\mathbf{U}$, and which does not lie too far away from the body, then the integrand in (4.6.3) can be replaced by its quadratic approximation, on using (1.11.17) for the pressure, and (1.11.21) for $\rho\mathbf{u}$. This gives

$$\mathbf{F}=\rho_0\int_{S'} \{(\mathbf{U}.\mathbf{v}+\tfrac{1}{2}\mathbf{v}.\mathbf{w})\mathbf{n}-\mathbf{v}(\mathbf{U}+\mathbf{w}).\mathbf{n}\}\,\mathrm{d}S, \tag{4.6.4}$$

approximately, on using the result

$$\rho_0\int_{S'} (\mathbf{U}+\mathbf{w}).\mathbf{n}\,\mathrm{d}S=0, \tag{4.6.5}$$

which expresses the conservation of mass inside S'. The integral in (4.6.4) is invariant for changes of S', such that S' always encloses the same singularities of $\mathbf{v}$, when $\mathbf{v}$ and $\mathbf{w}$ satisfy the linearized equations for irrotational flow with no sources, as can be shown by using the identity (A. 1.2). But when there are vortex sheets in the flow, and the velocity is singular at their edges, (4.6.4) is not invariant for all changes of S'; in fact, the assumptions made in its derivation are violated since $\mathbf{v}$ is not then uniformly small on S', and an impasse appears to have been reached. This point was argued at length by Ursell and Ward (1950), and it appears that, without a more exact knowledge of the behaviour of the vorticity in the vortex sheet, the difficulty can only be overcome by making a further assumption. This unsatisfactory state of affairs seems to have been overlooked by previous writers. The assumption made in the paper cited is equivalent to the assumption that S' cannot be entirely arbitrary, but must belong to the set of surfaces, S^*, which enclose the body, and *pass through its trailing edge.* In this case, (4.6.4) gives both the first-order lateral force, perpendicular to $\mathbf{U}$, and the second-order drag force, parallel to $\mathbf{U}$, correctly always, and gives the lateral force of the second order correctly in special cases. The justification of these remarks is made below.

If the particle velocity $\mathbf{u}$ is expanded in a series of the form

$$\mathbf{u}=\mathbf{U}+\mathbf{v}_{\mathrm{I}}+\mathbf{v}_{\mathrm{II}}+\mathbf{v}_{\mathrm{III}}+\ldots, \tag{4.6.6}$$

where $\mathbf{v}_{\mathrm{I}}$, $\mathbf{v}_{\mathrm{II}}$, $\mathbf{v}_{\mathrm{III}}$, etc., are of order t, t^2, t^3, etc., and t is the thickness parameter for the body in question, then $\mathbf{v}_{\mathrm{I}}$ is the linearized perturbation velocity, denoted by $\mathbf{v}$ previously, and it has a potential ϕ_{I}. Since the entropy changes through shocks are of order t^3, the second-order perturbation velocity $\mathbf{v}_{\mathrm{II}}$ has a potential ϕ_{II}, but in supersonic flow, $\mathbf{v}_{\mathrm{III}}$, etc., are not irrotational vector fields; however, the terms of third and higher order are not required in what follows.

The correct second approximation to Bernoulli's equation is

$$(p-p_0)/\rho_0=-\mathbf{U}.(\mathbf{v}_{\mathrm{I}}+\mathbf{v}_{\mathrm{II}})-\tfrac{1}{2}\mathbf{v}_{\mathrm{I}}.\mathbf{w}_{\mathrm{I}}, \tag{4.6.7}$$

instead of the quadratic approximation used when deriving (4.6.4) (cf. the remarks after (1.11.17)). Hence the integrand in (4.6.4) requires amendment by the addition of terms $\mathbf{U}.\mathbf{v}_{\mathrm{II}}\mathbf{n}-\mathbf{v}_{\mathrm{II}}\mathbf{U}.\mathbf{n}$. With this amendment, the aerodynamic force can be expanded as

$$\mathbf{F}=\mathbf{F}_{\mathrm{I}}+\mathbf{F}_{\mathrm{II}}+\ldots, \tag{4.6.8}$$

$\mathbf{F}_{\mathrm{I}}$ and $\mathbf{F}_{\mathrm{II}}$ being respectively the terms of order t and t^2 in $\mathbf{F}$, and where

$$\mathbf{F}_{\mathrm{I}}=\rho_0\int_{S'}(\mathbf{U}.\mathbf{v}_{\mathrm{I}}\mathbf{n}-\mathbf{v}_{\mathrm{I}}\mathbf{U}.\mathbf{n})\,\mathrm{d}S=\rho_0\mathbf{U}\wedge\int_{S'}\mathbf{n}\wedge\mathbf{v}_{\mathrm{I}}\,\mathrm{d}S, \tag{4.6.9}$$

$$\mathbf{F}_{\mathrm{II}}=\rho_0\int_{S'}(\tfrac{1}{2}\mathbf{v}_{\mathrm{I}}.\mathbf{w}_{\mathrm{I}}\mathbf{n}-\mathbf{v}_{\mathrm{I}}\mathbf{w}_{\mathrm{I}}.\mathbf{n})\,\mathrm{d}S+\rho_0\mathbf{U}\wedge\int_{S'}\mathbf{n}\wedge\mathbf{v}_{\mathrm{II}}\,\mathrm{d}S. \tag{4.6.10}$$

The first-order force $\mathbf{F}_{\mathrm{I}}$ is always perpendicular to the stream direction. The integral over S' is invariant for all changes of S' such that S' encloses the body, even when linearized vortex sheets are present, as can be seen by putting $\mathbf{v}_{\mathrm{I}}=\nabla\phi_{\mathrm{I}}$, in which case

$$\int_{S'}\mathbf{n}\wedge\mathbf{v}_{\mathrm{I}}\,\mathrm{d}S=\int_{L'}\phi_{\mathrm{I}}\,\mathrm{d}\mathbf{s}, \tag{4.6.11}$$

where L' is both sides of the curve in which the linearized vortex sheet intersects S'. When there are no vortex sheets, (4.6.11) shows that $\mathbf{F}_{\mathrm{I}}=0$. The drag force must be given from $\mathbf{F}_{\mathrm{II}}$ and must always be a second-order force at most. The second-order perturbation velocity $\mathbf{v}_{\mathrm{II}}$ does not contribute to the second-order drag force, as (4.6.10) shows. The first integral in (4.6.10) is not invariant for all changes of S' such that S' encloses the body, since integrals of this form taken over two surfaces S' and S'' differ by the value of a similar integral taken over both sides and the edges of that part of the linearized vortex sheet which lies between S' and S''. The contributions from both sides of the vortex sheet cancel, but the integrals over the edges (which may be evaluated by enclosing the edges in small circular cylinders whose radii are allowed to tend to zero) give contributions which are perpendicular to the edges, and hence to $\mathbf{U}$, as is shown in §4.7. Thus the drag force obtainable from $\mathbf{F}_{\mathrm{II}}$ is invariant for changes of S' when vortex sheets are present.

If the series (4.6.8) converges for small t, then $\mathbf{F}_{\mathrm{I}}$ gives the first-order force correctly, and the drag component of $\mathbf{F}_{\mathrm{II}}$ should be correct, since they both depend only on the linearized term $\mathbf{v}_{\mathrm{I}}$, and their defining integrals are invariant with respect to changes of the surface S'. The lateral component of $\mathbf{F}_{\mathrm{II}}$ is not correct as written, and it is probable that the reason for this is to be found in the non-uniform convergence of the series (4.6.6) at the edges of the vortex sheet.

Now take S' to be the body surface, Σ^*, plus small tube-like surfaces of ultimately vanishing radius which enclose the edges at which $\mathbf{u}$ is singular. Then, correct to the second order, the amended version of (4.6.4) gives

$$\mathbf{F}=\rho_0\int_{\Sigma^*}\{\mathbf{U}.(\mathbf{v}_{\mathrm{I}}+\mathbf{v}_{\mathrm{II}})+\tfrac{1}{2}\mathbf{v}_{\mathrm{I}}.\mathbf{w}_{\mathrm{I}}\}\mathbf{n}\,dS+\mathbf{F}_e, \qquad (4.6.12)$$

where $\mathbf{F}_e$ is the total force on the edges, arising as the contributions from the tube-like surfaces, and is a second-order term. Since it has been shown that the first-order lateral force and the second-order drag force are invariant for changes in S', it follows that they are given correctly by (4.6.12); in fact, for thin bodies, these forces are given correctly by the lower approximation

$$\mathbf{F}=\rho_0\int_{\Sigma^*}\mathbf{U}.\mathbf{v}_{\mathrm{I}}\mathbf{n}\,dS+\mathbf{F}_e, \qquad (4.6.13)$$

since the terms omitted give a second-order lateral force and a third-order drag force, because $\mathbf{U}.\mathbf{n}$ is a quantity of order t. The results (4.6.12) or (4.6.13) show that these forces can be calculated correctly by integrating the approximate normal pressure over the body surface, and adding the appropriate edge suction force, even though the approximation to the pressure is not a valid form near the edges where $\mathbf{u}$ is infinite. Since the drag is given correctly by this process, it seems likely that the second-order lateral force is also given by it, and hence that (4.6.12) gives the correct overall force up to and including terms of the second order. If S^* is any surface which encloses the body and passes through the trailing edge, then (4.6.12) is equivalent to

$$\mathbf{F}=\rho_0\int_{S^*}\{[\mathbf{U}.(\mathbf{v}_{\mathrm{I}}+\mathbf{v}_{\mathrm{II}})+\tfrac{1}{2}\mathbf{v}_{\mathrm{I}}.\mathbf{w}_{\mathrm{I}}]\mathbf{n}-(\mathbf{v}_{\mathrm{I}}+\mathbf{v}_{\mathrm{II}})\mathbf{U}.\mathbf{n}-\mathbf{v}_{\mathrm{I}}\mathbf{w}_{\mathrm{I}}.\mathbf{n}\}\,dS, \qquad (4.6.14)$$

correct to terms of the second order. This can be written

$$\mathbf{F}=\rho_0\mathbf{U}\wedge\int_{L^*}\phi_{\mathrm{I}}\,d\mathbf{s}+\rho_0\int_{S^*}(\tfrac{1}{2}\mathbf{v}_{\mathrm{I}}.\mathbf{w}_{\mathrm{I}}\mathbf{n}-\mathbf{v}_{\mathrm{I}}\mathbf{w}_{\mathrm{I}}.\mathbf{n})\,dS+\rho_0\mathbf{U}\wedge\int_{L^*}\phi_{\mathrm{II}}\,d\mathbf{s}, \qquad (4.6.15)$$

where L^* is both sides of the trailing edge where ϕ_{I} and ϕ_{II} are discontinuous. In general the term in ϕ_{II} is of the second order, and then the second-order lateral force cannot be calculated by linearized theory alone. For some special bodies, however, the component of the term in ϕ_{II} which is perpendicular to the first-order force (given by the term in ϕ_{I}) can be shown to be of the third order, and the lateral force in this direction is then a second-order force that can be calculated from linearized theory alone. Such a case is that of a nearly plane wing, for which the component of ds on L^* normal to the mean plane is of the first order, in which case the lateral force parallel to the mean plane is given correctly to the second order by linearized theory.

A similar argument for slender bodies leads to the result (4.6.15); thus for all bodies, the aerodynamic force can be determined from this relation. The above analysis shows that continuity considerations require the approximate aerodynamic force to be defined by (4.6.15), and all alternative expressions must be derived from this form.

For thin bodies, if Σ is both sides of the mean surface, with outward normal $\boldsymbol{\nu}$ on each side, the most generally useful of the alternative expressions are for the first-order lateral force, $\mathbf{F}_{\text{I}}$,

$$\mathbf{F}_{\text{I}} = \rho_0 \int_{\Sigma} \mathbf{U}.\mathbf{v}\boldsymbol{\nu}\,dS = -\int_{\Sigma} (p - p_0)\boldsymbol{\nu}\,dS, \qquad (4.6.16)$$

and for the drag force, D,

$$D = \mathbf{k}.\mathbf{F}_{\text{II}} = \rho_0 \int_{\Sigma} \mathbf{U}.\mathbf{v}\mathbf{k}.\mathbf{n}\,dS + \mathbf{k}.\mathbf{F}_e = -\mathbf{k}.\int_{\Sigma} (p - p_0)\mathbf{n}\,dS + \mathbf{k}.\mathbf{F}_e, \qquad (4.6.17)$$

where $\mathbf{k}$ is the unit vector in the direction of $\mathbf{U}$, and $\mathbf{n}$ is the outward normal to the actual body surface. These two equations also give the forces when $\mathbf{v}$ is replaced by $\mathbf{v}'$ as defined in § 1.12.

4.7. Edge forces

It is shown in § 4.6 that, when evaluating the second-order forces by integrating pressure forces over the mean body surface, care must be taken to include the forces on edges at which the velocity is infinite in linearized theory. These forces are of two distinct types. The first type can arise at a sharp edge for which the slope of the body surface is bounded, and corresponds to the kind of singularity that occurs on leading edges of lifting aerofoils for example. This type of singularity is inherent in the linearized equations (Lighthill, 1951). The second type arises from the infinite surface slope at rounded leading edges that, strictly speaking, are outside the scope of linearized theory. It has been shown by R. T. Jones (1950) that linearized theory nevertheless gives the correct forces in this case provided that the edge forces are taken into account.

To obtain expressions for the edge forces, take local rectangular cartesian co-ordinates x', y', z' with origin at the edge for the point of interest, such that x' is measured outwards normally from the edge, the y'-axis is normal to the local tangent plane to the mean body surface, and the z'-axis is tangent to the edge. Then, if χ is the angle of sweep of the edge, the approximate equation for ϕ in a neighbourhood of an edge at which the velocity is infinite is

$$(1-M^2\cos^2\chi)\frac{\partial^2\phi}{\partial x'^2}+\frac{\partial^2\phi}{\partial y'^2}=0, \qquad (4.7.1)$$

since the variation of ϕ with z' is small compared with its variations with x' and y' near the edge. For singularities of the first type mentioned above, when the normal velocity on the mean surface is bounded, the only singular solution of (4.7.1) which gives a finite non-zero edge force and for which ϕ is bounded is

$$\frac{\partial\phi}{\partial x'}=\Im\frac{K}{\beta'\sqrt{(x'+i\beta'y')}}, \quad \frac{\partial\phi}{\partial y'}=\Re\frac{K}{\sqrt{(x'+i\beta'y')}}, \qquad (4.7.2)$$

where $\beta'^2=1-M^2\cos^2\chi$, and K is a function of z' only. In any particular problem K can usually be determined most easily from the behaviour of the velocity component $\partial\phi/\partial y'$ just outside the body in the plane $y'=0$:

$$K=\lim_{x'\to+0}\sqrt{x'}\left(\frac{\partial\phi}{\partial y'}\right)_{y'=0}. \qquad (4.7.3)$$

A simple method of evaluating the edge force, which for (4.7.2) turns out to be a suction force on the edge, is to take the surface of integration in the integral

$$\rho_0\int_S(\tfrac{1}{2}\mathbf{v}.\mathbf{wn}-\mathbf{vw}.\mathbf{n})\,dS \qquad (4.7.4)$$

to consist of two surfaces $y'=\pm\epsilon$, $x'\leqslant\delta$ say, and a surface $x'=\delta$, $-\epsilon<y'<\epsilon$, which enclose the edge, and to let $\epsilon\to 0$ for fixed δ. Then the x'-component of the suction force per unit length of edge is

$$\begin{aligned}
&\lim_{\epsilon\to 0} 2\rho_0\int_{-\delta}^{\delta}\left(\frac{\partial\phi}{\partial x'}\frac{\partial\phi}{\partial y'}\right)_{y'=\epsilon}dx' \\
&=\lim_{\epsilon\to 0}\frac{2\rho_0}{\beta'}\int_{-\delta}^{\delta}\Im\left(\frac{K}{\sqrt{(x'+i\beta'\epsilon)}}\right)\Re\left(\frac{K}{\sqrt{(x'+i\beta'\epsilon)}}\right)dx' \\
&=\frac{\rho_0K^2}{\beta'}\lim_{\epsilon\to 0}\int_{-\delta}^{\delta}\Im\left(\frac{1}{x'+i\beta'\epsilon}\right)dx' \\
&=\frac{\rho_0K^2}{\beta'}\lim_{\epsilon\to 0}\int_{-\delta}^{\delta}\frac{\beta'\epsilon}{x'^2+\beta'^2\epsilon^2}\,dx'=\frac{\pi\rho_0K^2}{\sqrt{(1-M^2\cos^2\chi)}}, \qquad (4.7.5)
\end{aligned}$$

and the other components vanish.† This force occurs only on subsonic edges, where $1-M^2\cos^2\chi>0$.

In the above evaluation of the suction force, the lower limit is arbitrary and is written as $-\delta$ for convenience; and the contribution from the surface $x'=\delta$, $-\epsilon<y'<\epsilon$ vanishes in the limit.

For the case of a rounded leading edge, with radius of curvature r_0, the solution required is of similar form to (4.7.2) but with K replaced by $ir_0^{\frac{1}{2}}U\cos\chi$, and the edge force is then of opposite sense to (4.7.5) and of magnitude (R.T. Jones, 1950)

$$\frac{\pi\rho_0 r_0 U^2\cos^2\chi}{\sqrt{(1-M^2\cos^2\chi)}}. \tag{4.7.6}$$

Again this force occurs only on subsonic edges.

When both types of singularity occur together, the cross-terms between them cancel, and the resultant force on the edge is obtained by combining (4.7.5) and (4.7.6) with appropriate senses.

4.8. Aerodynamic moments

The moment of the pressure forces on a body can be found either by direct integration over the body surface or by considering the flux of angular momentum through a closed surface surrounding the body. The difficulties and arguments involved are similar to those given in §4.6 for the aerodynamic force, and need not be repeated.

If $\mathbf{R}_1$ is the position vector of the point about which the vector moment, $\mathbf{T}$, is required, and $\mathbf{R}$ is the position vector of the surface element in the integrals, then $\mathbf{T}$ is defined by

$$\begin{aligned}\mathbf{T}=\rho_0\int_{S^*}&(\mathbf{R}-\mathbf{R}_1)\wedge(\mathbf{U}.\mathbf{v}_\mathrm{I}\mathbf{n}-\mathbf{v}_\mathrm{I}\mathbf{U}.\mathbf{n}-\mathbf{U}\mathbf{w}_\mathrm{I}.\mathbf{n})\,dS\\ +\rho_0\int_{S^*}&(\mathbf{R}-\mathbf{R}_1)\wedge(\tfrac{1}{2}\mathbf{v}_\mathrm{I}.\mathbf{w}_\mathrm{I}\mathbf{n}-\mathbf{v}_\mathrm{I}\mathbf{w}_\mathrm{I}.\mathbf{n})\,dS\\ +\rho_0\int_{S^*}&(\mathbf{R}-\mathbf{R}_1)\wedge(\mathbf{U}.\mathbf{v}_\mathrm{II}\mathbf{n}-\mathbf{v}_\mathrm{II}\mathbf{U}.\mathbf{n}-\mathbf{U}\mathbf{w}_\mathrm{II}.\mathbf{n})\,dS,\end{aligned} \tag{4.8.1}$$

correct to terms of the second order for both thin and slender bodies. As before, S^* is any surface which encloses the body and passes through the trailing edge.

† The author is indebted to Professor M. J. Lighthill for this elegant analysis.

Usually only the first-order moment is required. For thin bodies, this is

$$\mathbf{T}_{\mathrm{I}}=\rho_0\int_\Sigma \mathbf{U}.\mathbf{v}(\mathbf{R}-\mathbf{R}_1)\wedge\boldsymbol{\nu}\,\mathrm{d}S=-\int_\Sigma (p-p_0)(\mathbf{R}-\mathbf{R}_1)\wedge\boldsymbol{\nu}\,\mathrm{d}S, \tag{4.8.2}$$

where Σ is both sides of the mean body surface, and $\boldsymbol{\nu}$ is its outward normal on each side (cf. (4.6.16)).

4.9. Uniqueness

It can now be shown that there is, at most, one solution of the linearized equations

$$\nabla\wedge\mathbf{v}=0,\quad \nabla.\mathbf{w}=0,\quad \mathbf{w}=\boldsymbol{\Psi}.\mathbf{v}, \tag{4.9.1}$$

which satisfies the boundary conditions of § 4.5 for thin bodies of finite dimensions.† Let $\mathbf{v}_1$ and $\mathbf{v}_2$ be any two solutions, different if this is possible, and let $\mathbf{v}=\mathbf{v}_1-\mathbf{v}_2$. Then $\mathbf{v}$ also satisfies (4.9.1), the boundary conditions $\boldsymbol{\nu}.\mathbf{v}=0$ on the mean body surface Σ, $\mathbf{U}.\mathbf{v}=0$ on free boundaries, and conditions (iii)–(ix) of § 4.5.

In the identity (A. 1.2) multiplied scalarly by $\mathbf{U}$, let $\mathbf{v}=\mathbf{v}_1-\mathbf{v}_2$, $\mathbf{w}=\mathbf{w}_1-\mathbf{w}_2$. Then this identity becomes

$$\int_S (\mathbf{U}.\mathbf{v}\mathbf{w}.\mathbf{n}-\tfrac{1}{2}\mathbf{v}.\mathbf{w}\mathbf{U}.\mathbf{n})\,\mathrm{d}S=0, \tag{4.9.2}$$

by virtue of (4.9.1). The proof of uniqueness is now in two parts, (i) for subsonic flow, and (ii) for supersonic flow.

(i) *Subsonic flow.* Take the closed surface S in (4.9.2) to consist of five parts:

Σ: both sides of the mean body surface, plus small tube-like surfaces surrounding its edges,

σ: both sides of the linearized vortex sheet, plus small tube-like surfaces surrounding its edges (if any),

S_1: a cylindrical surface of large radius R_0, whose axis is parallel to $\mathbf{U}$ and passes near to the body,

S_2: a plane $\mathbf{R}.\mathbf{U}=-R_0U$, at a large distance R_0 upstream from the body,

S_3: a plane $\mathbf{R}.\mathbf{U}=R_0U$, at a large distance R_0 downstream from the body.

Now on the mean body surface, with normal $\boldsymbol{\nu}$, $\boldsymbol{\nu}.\mathbf{w}=\boldsymbol{\nu}.\mathbf{v}=0$ and $\mathbf{U}.\boldsymbol{\nu}=0$, so the only contribution from the surface integral over

† Ursell and Ward (1950).

Σ comes from the tube-like surfaces which enclose the leading edge, on which $\mathbf{v}$ may be infinite. This is of the form

$$-\pi U\int\frac{K^2|\cos\chi\,ds|}{\sqrt{(1-M^2\cos^2\chi)}},\tag{4.9.3}$$

where the integration is along the leading edge, from the results of §4.7; forces like (4.7.6) give no contribution because the leading edge is not rounded in the difference flow.

The integral over σ vanishes, since the contributions from the opposite sides of the vortex sheet cancel, and the contribution from the tube-like surfaces to the momentum integral in the form (4.7.4) is perpendicular to $\mathbf{U}$, and vanishes on scalar multiplication by $\mathbf{U}$.

The integrals over S_1 and S_2 are $O(R_0^{-2})$ as $R_0\to\infty$, from condition (viii) of §4.5, and this same condition gives the value of the integral over S_3 as

$$-\tfrac{1}{2}U\int_{S_3}(u^2+v^2)\,dS+O(R_0^{-2})\tag{4.9.4}$$

as $R_0\to\infty$, u, v, w being the rectangular components of $\mathbf{v}$, with w as the component parallel to $\mathbf{U}$ as usual.

On proceeding to the limit, $R_0\to\infty$, (4.9.2) gives

$$-\pi U\int\frac{K^2|\cos\chi\,ds|}{\sqrt{(1-M^2\cos^2\chi)}}-\tfrac{1}{2}U\int_{S_3}(u^2+v^2)\,dS=0.\tag{4.9.5}$$

Both integrals in (4.9.5) must converge by condition (vii) of §4.5, and, since the integrands are essentially positive,

$$u=v=0 \text{ on } S_3 \quad\text{and}\quad K=0.\tag{4.9.6}$$

Therefore $\mathbf{v}$ is continuous on S_3 and there is no vortex sheet in the difference flow; hence $\mathbf{v}$ is continuous and bounded everywhere, except possibly at leading edges where $|r^{\frac{1}{2}}\mathbf{v}|$ must tend to zero as $r\to 0$, r being distance from the edge.

An application of the identity (2.4.3),

$$\int_S\phi\mathbf{w}.\mathbf{n}\,dS=\int_V(u^2+v^2+\beta^2w^2)\,dV,\tag{4.9.7}$$

to the space between the body and a sphere of large radius R_0, now shows that $\mathbf{v}=0$ everywhere, just as for the uniqueness result of §2.4. Hence $\mathbf{v}_1=\mathbf{v}_2$, and there can be at most one solution for subsonic flow.

The presence of free boundaries modifies the proof only very slightly. Part of the closed surface S in (4.9.2) must now be the mean surface of the free boundary, and its contribution to (4.9.2) vanishes, since $\mathbf{U}.\mathbf{v}$ and $\mathbf{U}.\mathbf{n}$ both vanish on this surface. The proof then follows the above lines.

(ii) *Supersonic flow.* Take the closed surface S in (4.9.2) to consist of Σ, σ and S_3, as in the subsonic case, plus the characteristic surface, S_4, that bounds the influence domain of the body.

The contributions from Σ and σ are the same as before, and the contribution from S_3 for finite R_0 (not necessarily large here) is

$$-\tfrac{1}{2}U\int_{S_3}(u^2+v^2+B^2w^2)\,\mathrm{d}S. \tag{4.9.8}$$

The contribution from S_4 vanishes since $\mathbf{v}_1$, $\mathbf{v}_2$ and $\mathbf{v}$ vanish just upstream from S_4, and the integrand in (4.9.2) is proved in Appendix 1 to be continuous at a characteristic surface.

The identity (4.9.2) now gives

$$-\pi U\int\frac{K^2\,|\cos\chi\,\mathrm{d}s\,|}{\sqrt{(1-M^2\cos^2\chi)}}-\tfrac{1}{2}U\int_{S_3}(u^2+v^2+B^2w^2)\,\mathrm{d}S=0, \tag{4.9.9}$$

the integration being over the leading subsonic edge in the first integral. Again both integrals converge by condition (vii) of §4.5, and, since the integrands are essentially positive,

$$u=v=w=0 \text{ on } S_3 \quad\text{and}\quad K=0. \tag{4.9.10}$$

The first of these is true for any S_3 that lies downstream from the body; hence ϕ vanishes over all space downstream from any S_3, because ϕ is continuous and is zero upstream from S_4. The uniqueness result in §3.2, for a properly set Cauchy initial value problem for the space upstream from any S_3, now shows that ϕ is zero in all space upstream from S_3. Therefore $\mathbf{v}=0$ everywhere, which proves uniqueness in the supersonic case.

The modification necessary when free boundaries are present is the same as for the subsonic case.

As a corollary, if $\mathbf{v}_1$ and $\mathbf{v}_2$ satisfy

$$\nabla\wedge\mathbf{v}=\boldsymbol{\zeta},\quad \nabla.\mathbf{w}=Q,\quad \mathbf{w}=\boldsymbol{\Psi}.\mathbf{v}, \tag{4.9.11}$$

where $\boldsymbol{\zeta}$ and Q are given, then the difference velocity $\mathbf{v}=\mathbf{v}_1-\mathbf{v}_2$ satisfies (4.9.1) and the same boundary conditions as before. Hence the above arguments prove uniqueness for the solution of (4.9.11).

Also the condition of finite dimensions for the body can be relaxed in numerous cases; for example, when the body is cylindrical at great distances upstream or downstream or both, or when the flow is two-dimensional.

The physical interpretation of the arguments leading up to (4.9.6) and (4.9.10) is not without interest. By comparing (4.9.2) with (4.6.15), it is seen that the above method of proof effectively considers the drag force on the mean body surface due to the velocity field $\mathbf{U}+\mathbf{v}_1-\mathbf{v}_2$. Clearly there is no contribution to the drag from the mean surface, but there is a thrust or negative drag contribution from any singular velocities at the subsonic leading edge. However, the drag is an essentially positive quantity, given by the integrals over S_3. This is a contradiction, and hence there is no drag, which implies that there cannot be any vortex sheets in the difference flow. This result having been established, uniqueness then follows by standard methods.

4.10. A general flow-reversal theorem

The flow past a given body can be connected with a flow past another (or the same) body having the same mean surface in which the velocity at infinity is equal in magnitude but opposite in direction. A very general relation of this type was established by Ursell and Ward (1950) as follows.

Let $\mathbf{n}_1$ and $\mathbf{n}_2$ be vector functions of position on the surfaces of two thin bodies having the same mean surface, Σ, the two bodies being respectively in flows for which the undisturbed streams have velocities $\mathbf{U}$ and $-\mathbf{U}$, the same densities, ρ_0, and the same Mach number. Let $\mathbf{v}_1$ and $\mathbf{v}_2$ be perturbation velocities satisfying the equations of motion

$$\nabla \wedge \mathbf{v}=0, \quad \nabla . \mathbf{w}=0, \quad \mathbf{w}=\mathbf{\Psi}.\mathbf{v}, \tag{4.10.1}$$

satisfying the boundary conditions

$$\boldsymbol{\nu}.\mathbf{v}_1=-\mathbf{U}.\mathbf{n}_1, \quad \boldsymbol{\nu}.\mathbf{v}_2=\mathbf{U}.\mathbf{n}_2, \tag{4.10.2}$$

respectively on Σ, where $\boldsymbol{\nu}$ is the unit outward normal to Σ, and otherwise satisfying the boundary conditions (iii)–(ix) of §4.5. Let p_1 and p_2 be the corresponding linearized perturbation pressures:

$$p_1=-\rho_0\mathbf{U}.\mathbf{v}_1, \quad p_2=\rho_0\mathbf{U}.\mathbf{v}_2. \tag{4.10.3}$$

Then $$\mathbf{U}.\int_{\Sigma}(p_1\mathbf{n}_1+p_2\mathbf{n}_2)\,\mathrm{d}S=0. \tag{4.10.4}$$

This result can be established by using the identity (A.1.3) multiplied scalarly by $\rho_0\mathbf{U}$, which, by virtue of (4.10.1), becomes

$$\rho_0\int_S(\mathbf{U}.\mathbf{v}_1\mathbf{w}_2.\mathbf{n}+\mathbf{U}.\mathbf{v}_2\mathbf{w}_1.\mathbf{n}-\mathbf{v}_1.\mathbf{w}_2\mathbf{U}.\mathbf{n})\,\mathrm{d}S=0. \tag{4.10.5}$$

Take the closed surface, S, to consist of both sides of the mean body surface, Σ, the surfaces of any linearized vortex sheets, the surface at infinity, and tube-like surfaces of ultimately vanishing radius surrounding the edges of Σ and the vortex sheets.

The contribution from the surface at infinity vanishes; in subsonic flow, as a consequence of condition (viii), §4.5, and, in supersonic flow, as a consequence of condition (ix), §4.5, since the integrand of (4.10.5) is proved in Appendix 1 to be continuous on characteristic surfaces. The contribution from the tube-like surfaces at the edges of Σ vanishes, since $\mathbf{v}_1$ is finite where $\mathbf{v}_2$ is singular, and vice versa, from condition (iii), §4.5. The integrals over the tube-like surfaces at the edges of the vortex sheets also vanish, since any 'suction forces' there are perpendicular to $\mathbf{U}$. Hence the only contribution to the surface integral over S comes from Σ, and (4.10.5) becomes

$$\rho_0\int_{\Sigma}(\mathbf{U}.\mathbf{v}_1\mathbf{w}_2.\boldsymbol{\nu}+\mathbf{U}.\mathbf{v}_2\mathbf{w}_1.\boldsymbol{\nu}-\mathbf{v}_1.\mathbf{w}_2\mathbf{U}.\boldsymbol{\nu})\,\mathrm{d}S=0, \tag{4.10.6}$$

since $\mathbf{n}=-\boldsymbol{\nu}$ on Σ.

Now, on Σ, $\mathbf{U}.\boldsymbol{\nu}=0$ and $\mathbf{w}.\boldsymbol{\nu}=\mathbf{v}.\boldsymbol{\nu}$. Hence, by using (4.10.2) and (4.10.3), the identity (4.10.6) gives (4.10.4), which proves the theorem.

As a corollary it may be noticed that the theorem is still true if the flows include free boundaries on whose mean surfaces $\mathbf{U}.\mathbf{v}_1$ and $\mathbf{U}.\mathbf{v}_2$ vanish.

The theorem is also true in numerous cases when the body dimensions are not finite; for example, when the flow is two-dimensional, or when the body is cylindrical at great distances upstream or downstream or both (Flax, 1952*b*), or when the linearized free boundaries are of infinite extent.

4.11. Some special cases of the general flow-reversal theorem

The significance of the general theorem is difficult to appreciate in the form given in §4.10, and some special cases of more immediate practical application are derived below. Many of these special cases have been derived for plane wings by previous writers.†

A derivation of exceptional interest has been given by Brown (1950), who generalizes a method, due to Munk (1950), which involves an almost entirely physical argument substantially equivalent to the method of §4.10.

(i) The simplest special case of (4.10.4) occurs when the bodies in both flows are the same, and $\mathbf{n}_1 = \mathbf{n}_2 = \mathbf{n}$, where $\mathbf{n}$ is the unit outward normal to the body surface. Then (4.10.4) gives

$$\mathbf{U}.\int_\Sigma p_1 \mathbf{n}\, dS = -\mathbf{U}.\int_\Sigma p_2 \mathbf{n}\, dS. \qquad (4.11.1)$$

Comparison with (4.6.17) shows that this result means that the drag forces, neglecting any edge forces, are of the same magnitude in the two flows. This result was given by Hayes (1947) and von Kármán (1947) for the still more special case when Σ is a plane.

(ii) With the same body in both flows, as in (i), let a portion Σ_1 of the body surface be given an extra small vector incidence $\boldsymbol{\alpha}_1$ in the flow $\mathbf{U}$. Then if $\mathbf{n}$ was the normal to the original body surface, the new normal is approximately $\mathbf{n} + \boldsymbol{\alpha}_1 \wedge \boldsymbol{\nu}$, neglecting second-order quantities. Take

$$\left.\begin{aligned} &\mathbf{n}_1 = \boldsymbol{\alpha}_1 \wedge \boldsymbol{\nu} \text{ on } \Sigma_1, \\ &\mathbf{n}_1 = 0 \text{ on the rest of } \Sigma; \\ &\mathbf{n}_2 = \mathbf{n} \text{ on } \Sigma. \end{aligned}\right\} \qquad (4.11.2)$$

Then p_1 is the excess pressure consequent upon the change in the body surface, and (4.10.4) gives

$$\mathbf{U}.\int_\Sigma p_1 \mathbf{n}\, dS = -\mathbf{U}.\int_{\Sigma_1} p_2 \boldsymbol{\alpha}_1 \wedge \boldsymbol{\nu}\, dS \qquad (4.11.3)$$

or

$$-\mathbf{k}.\int_\Sigma p_1 \mathbf{n}\, dS = \mathbf{k} \wedge \boldsymbol{\alpha}_1 . \int_{\Sigma_1} p_2 \boldsymbol{\nu}\, dS, \qquad (4.11.4)$$

where $\mathbf{k}$ is the unit vector parallel to $\mathbf{U}$.

† C. E. Brown (1950), A. H. Flax (1949, 1952*a*), S. Harmon (1949*b*), W. D. Hayes (1947, 1948), T. von Kármán (1947), M. M. Munk (1950), G. N. Ward (1949*d*).

Comparison with (4.6.17) shows that the left-hand side of this equation is the extra drag force (neglecting edge forces) on the original body, due to the extra incidence of Σ_1, and comparison with (4.6.16) shows that the right-hand side is proportional to the component in the plane of incidence of the original lateral force on Σ_1 in the reversed flow. (The plane of incidence is a plane perpendicular to the vector $\boldsymbol{\alpha}_1$.)

(iii) Again with the same body in both flows, let a portion Σ_1 of the body surface be given a very small angular velocity $\boldsymbol{\Omega}_1$ about a point $\mathbf{R}_1$. If $\boldsymbol{\Omega}_1$ is small enough, the motion is quasi-steady, and it is usually assumed that the forces are given correctly in linearized theory by warping the body suitably and calculating the steady flow past the warped body. This assumption is adopted here without further justification. The boundary condition of zero relative normal velocity on the body surface at the point $\mathbf{R}$ is

$$\mathbf{n}.(\mathbf{U}+\mathbf{v})=\mathbf{n}.\boldsymbol{\Omega}_1\wedge(\mathbf{R}-\mathbf{R}_1), \qquad (4.11.5)$$

or, on the mean body surface, on neglecting second-order quantities,

$$\boldsymbol{\nu}.\mathbf{v}=-\mathbf{U}.\mathbf{n}+\boldsymbol{\nu}.\boldsymbol{\Omega}_1\wedge(\mathbf{R}-\mathbf{R}_1). \qquad (4.11.6)$$

In this case, take

$$\left.\begin{aligned}\boldsymbol{\nu}.\mathbf{v}_1=-\mathbf{U}.\mathbf{n}_1=\boldsymbol{\nu}.\boldsymbol{\Omega}_1\wedge(\mathbf{R}-\mathbf{R}_1)\text{ on }\Sigma_1,\\ \mathbf{n}_1=0\text{ on the rest of }\Sigma;\\ \mathbf{n}_2=\mathbf{n}\text{ on }\Sigma.\end{aligned}\right\} \qquad (4.11.7)$$

Then (4.10.4) gives

$$\mathbf{U}.\int_\Sigma p_1\mathbf{n}\,\mathrm{d}S=\int_{\Sigma_1}p_2\boldsymbol{\nu}.\boldsymbol{\Omega}_1\wedge(\mathbf{R}-\mathbf{R}_1)\,\mathrm{d}S, \qquad (4.11.8)$$

or

$$-\mathbf{k}.\int_\Sigma p_1\mathbf{n}\,\mathrm{d}S=-\frac{\boldsymbol{\Omega}_1}{U}.\int_{\Sigma_1}p_2(\mathbf{R}-\mathbf{R}_1)\wedge\boldsymbol{\nu}\,\mathrm{d}S. \qquad (4.11.9)$$

The left-hand side is the drag force in the original flow caused by the rotation of Σ_1, and, from (4.8.2), the right-hand side is proportional to the component, parallel to the axis of rotation, of the moment about the point $\mathbf{R}_1$ of the original pressures on Σ_1 in the reversed flow.

(iv) With any two bodies having the same mean surface, Σ, let a part, Σ_1, of the body surface in the stream $\mathbf{U}$ be given a small extra vector incidence $\boldsymbol{\alpha}_1$, and a part, Σ_2, of the body surface in the stream $-\mathbf{U}$ be given a small extra vector incidence $\boldsymbol{\alpha}_2$. Then take (cf. case (ii))

$$\left.\begin{aligned} \mathbf{n}_1 &= \boldsymbol{\alpha}_1 \wedge \boldsymbol{\nu} \text{ on } \Sigma_1,\\ \mathbf{n}_1 &= 0 \text{ on the rest of } \Sigma;\\ \mathbf{n}_2 &= \boldsymbol{\alpha}_2 \wedge \boldsymbol{\nu} \text{ on } \Sigma_2,\\ \mathbf{n}_2 &= 0 \text{ on the rest of } \Sigma; \end{aligned}\right\} \quad (4.11.10)$$

so that p_1, p_2 are the pressure changes consequent upon the surface changes, and (4.10.4) gives

$$\mathbf{U} \wedge \boldsymbol{\alpha}_2 . \int_{\Sigma_2} p_1 \boldsymbol{\nu} \, dS = -\mathbf{U} \wedge \boldsymbol{\alpha}_1 . \int_{\Sigma_1} p_2 \boldsymbol{\nu} \, dS. \quad (4.11.11)$$

The left-hand side of this equation is proportional to the component in the plane of incidence for $\boldsymbol{\alpha}_2$, of the extra lateral force on the area Σ_2 due to the change of incidence on Σ_1 in the flow $\mathbf{U}$, and similarly for the right-hand side.

For the more specialized case when $\Sigma_1 = \Sigma_2 = \Sigma$, and $\boldsymbol{\alpha}_1$, $\boldsymbol{\alpha}_2$ are equal vectors perpendicular to $\mathbf{U}$, (4.11.11) shows that the slope of the curve of lift against incidence for a given body remains unaltered when the direction of the stream is reversed. This result was given by Brown (1950) for the special case when Σ is a plane (i.e. for a plane wing), and by other authors for special shapes of plane wings in supersonic flow only.

(v) As in (iv), let the area Σ_1 in the stream $\mathbf{U}$ be given a small extra vector incidence $\boldsymbol{\alpha}_1$, and let the area Σ_2 in the stream $-\mathbf{U}$ be given a small angular velocity $\boldsymbol{\Omega}_2$ about the point $\mathbf{R}_2$, and take

$$\left.\begin{aligned} \mathbf{n}_1 &= \boldsymbol{\alpha}_1 \wedge \boldsymbol{\nu} \text{ on } \Sigma_1,\\ \mathbf{n}_1 &= 0 \text{ on the rest of } \Sigma;\\ \mathbf{U} . \mathbf{n}_2 &= \boldsymbol{\nu} . \boldsymbol{\Omega}_2 \wedge (\mathbf{R} - \mathbf{R}_2) \text{ on } \Sigma_2,\\ \mathbf{n}_2 &= 0 \text{ on the rest of } \Sigma. \end{aligned}\right\} \quad (4.11.12)$$

Then p_1 and p_2 are the pressure changes consequent upon the surface changes, and (4.10.4) gives

$$-\boldsymbol{\Omega}_2 . \int_{\Sigma_2} p_1 (\mathbf{R} - \mathbf{R}_2) \wedge \boldsymbol{\nu} \, dS = \mathbf{U} \wedge \boldsymbol{\alpha}_1 . \int_{\Sigma_1} p_2 \boldsymbol{\nu} \, dS. \quad (4.11.13)$$

The interpretations of the two sides of this equation follow the same lines as in the previous cases. For the case when $\Sigma_1 = \Sigma_2 = \Sigma$, and $\boldsymbol{\alpha}_1$ and $\boldsymbol{\Omega}_2$ are parallel vectors perpendicular to $\mathbf{U}$, (4.11.13) shows that the extra pitching moment per unit incidence in the stream $\mathbf{U}$ for a given body has the same magnitude as U times the extra lift per unit rate of pitch in the reversed flow. This result was also given by Brown (1950).

(vi) Let the area Σ_1 in the stream $\mathbf{U}$ be given a small angular velocity $\boldsymbol{\Omega}_1$ about the point $\mathbf{R}_1$; let the area Σ_2 in the stream $-\mathbf{U}$ be given a small angular velocity $\boldsymbol{\Omega}_2$ about the point $\mathbf{R}_2$, and take $\mathbf{n}_1$ as in (4.11.7) and $\mathbf{n}_2$ as in (4.11.12). Then (4.10.4) gives

$$\boldsymbol{\Omega}_2 . \int_{\Sigma_1} p_1(\mathbf{R}-\mathbf{R}_2)\wedge \boldsymbol{\nu}\, dS = \boldsymbol{\Omega}_1 . \int_{\Sigma_1} p_2(\mathbf{R}-\mathbf{R}_1)\wedge \boldsymbol{\nu}\, dS. \tag{4.11.14}$$

Special cases of this result were given by Brown (1950). For example, if $\boldsymbol{\Omega}_1$ is parallel to $\mathbf{U}$ and $\boldsymbol{\Omega}_2$ is perpendicular to $\mathbf{U}$, and $\Sigma_1 = \Sigma_2 = \Sigma$, then (4.11.14) shows that, for any thin body, the yawing (or pitching) moment due to a given rate of roll is equal to the rolling moment in the reversed flow due to the same rate of yaw (or pitch).

(vii) The six special cases given above are only illustrative, and are by no means exhaustive. They are not even independent, since some can be deduced from the others by superposition, a process that is clearly possible since p_1 and p_2 are linear functions of the perturbation velocities. New results can be obtained by superposition; for example, by combining results such as those in (ii), (iv) or (v) for different parts, Σ_1', Σ_1'' of Σ_1 with different incidences on each, it is possible to obtain flow-reversal results suitable for application to problems of aileron movement. Such examples can be multiplied indefinitely. The above results are also true for bodies in wind tunnels with either closed or open working sections of sufficient length, as can easily be seen, since part of Σ in the above can be the tunnel wall, and p_1 and p_2 will vanish at open sections of sufficient length.

4.12. A special flow-reversal theorem

In some special flows there are no vortex sheets, and no sufficiently singular velocities at edges to produce edge forces. In such cases, the aerodynamic force on the body producing the disturbance is

$$\mathbf{F} = \rho_0 \int_{\Sigma} (\mathbf{v}\mathbf{w}.\boldsymbol{\nu} - \tfrac{1}{2}\mathbf{v}.\mathbf{w}\boldsymbol{\nu})\, dS, \tag{4.12.1}$$

as follows from (4.6.15), Σ being the mean body surface.

Now if Σ is a plane or planes parallel to $y=0$ say, then the component of lateral force parallel to Σ is

$$F_x = \rho_0 \int_\Sigma u\mathbf{w}.\boldsymbol{\nu}\, dS. \tag{4.12.2}$$

The x-component of the identity (A.1.3), taken over the closed surface which consists of Σ and the surface at infinity, is

$$\int_\Sigma (u_1 \mathbf{w}_2.\boldsymbol{\nu} + u_2 \mathbf{w}_1.\boldsymbol{\nu})\, dS = 0, \tag{4.12.3}$$

since the integral over the surface at infinity vanishes. The boundary conditions on Σ for the two streams $\mathbf{U}$ and $-\mathbf{U}$ give

$$\boldsymbol{\nu}.\mathbf{w}_1 = \boldsymbol{\nu}.\mathbf{v}_1 = -\mathbf{U}.\mathbf{n} = -\boldsymbol{\nu}.\mathbf{v}_2 = -\boldsymbol{\nu}.\mathbf{w}_2, \tag{4.12.4}$$

so (4.12.3) becomes

$$\int_\Sigma (u_1 \mathbf{w}_1.\boldsymbol{\nu} + u_2 \mathbf{w}_2.\boldsymbol{\nu})\, dS = 0, \tag{4.12.5}$$

or, from (4.12.2), if there are no vortex sheets, etc., in the reversed flow

$$F_{1x} = -F_{2x}. \tag{4.12.6}$$

Thus the x-component of force is reversed in direction and is unchanged in magnitude. This result is trivial for subsonic flow since all the forces are then zero; its principal application is to plane symmetrical wings in supersonic flow.

An interesting consequence of (4.12.6) is that there is no side force on a nearly plane wing that is symmetrical about its chord plane and about a plane normal to the stream; for consideration of symmetry shows that $F_{1x} = F_{2x}$, and hence, from (4.12.6), $F_{1x} = F_{2x} = 0$.

4.13. Alternative general solutions for the potential

The general solution (2.6.10) of the linearized equation for the potential in irrotational flow suffers from the disadvantage that it cannot be applied directly to the solution of problems of subsonic flow past thin bodies, since ϕ and $\mathbf{w}.\mathbf{n}$ cannot be specified independently on the boundary. An adaptation can be made, however, which can be applied directly to certain types of problem.

In a subsonic flow past some body whose mean surface is given, let S_1, S_2, etc., be a finite number of cylindrical surfaces with generators parallel to $\mathbf{U}$, which extend upstream and downstream to infinity, and which, between them, contain the whole mean body surface. It follows that they contain also the linearized vortex sheets, if any, so they divide space into a finite number of connected domains V_1, V_2, etc., in each of which ϕ is regular. Now apply (2.6.10) to the domain V_1 bounded by a surface, S_1', consisting of those parts of S_1, S_2, etc., which bound V_1, and the relevant parts of the surface at infinity. If $\mathbf{R}_1$ is any finite point of V_1, then the integrals over the surface at infinity vanish by the boundary condition (viii) of §4.5, and if $\boldsymbol{\nu}_1$ is the outward normal to S_1', then $\mathbf{w}.\boldsymbol{\nu}_1 = \mathbf{v}.\boldsymbol{\nu}_1$ on S_1', and (2.6.10) gives

$$\phi(\mathbf{R}_1) = \frac{1}{4\pi}\int_{S_1'} \frac{\mathbf{v}(\mathbf{R}).\boldsymbol{\nu}_1}{R_\beta}\,\mathrm{d}S + \frac{\beta^2}{4\pi}\int_{S_1'} \frac{\phi(\mathbf{R})(\mathbf{R}-\mathbf{R}_1).\boldsymbol{\nu}_1}{R_\beta^3}\,\mathrm{d}S - \frac{1}{4\pi}\int_{V_1} \frac{Q(\mathbf{R})}{R_\beta}\,\mathrm{d}V. \quad (4.13.1)$$

The second integral can be integrated by parts with respect to z, and, with the lower limit of integration taken as infinity downstream for convenience, gives immediately

$$\frac{1}{4\pi}\int_{S_1'} \frac{\partial\phi}{\partial z}\frac{(\mathbf{R}-\mathbf{R}_1).\boldsymbol{\nu}_1}{(x-x_1)^2+(y-y_1)^2}\left(1+\frac{z_1-z}{R_\beta}\right)\mathrm{d}S, \quad (4.13.2)$$

the line integrals vanishing due to the boundary condition at infinity upstream and the choice of the lower limit. Hence (4.13.1) becomes

$$\phi(\mathbf{R}_1) = \frac{1}{4\pi}\int_{S_1'} \frac{\mathbf{v}(\mathbf{R}).\boldsymbol{\nu}_1}{R_\beta}\,\mathrm{d}S + \frac{1}{4\pi}\int_{S_1'} \mathbf{k}.\mathbf{v}(\mathbf{R})\frac{(\mathbf{R}-\mathbf{R}_1).\boldsymbol{\nu}_1}{(x-x_1)^2+(y-y_1)^2} \times\left(1+\frac{z_1-z}{R_\beta}\right)\mathrm{d}S - \frac{1}{4\pi}\int_{V_1}\frac{Q(\mathbf{R})}{R_\beta}\,\mathrm{d}V. \quad (4.13.3)$$

The process by which (2.6.10) was obtained in §2.6 can now be applied to the other domains V_2, V_3, etc., which do not include the point $\mathbf{R}_1$. This gives relations of the form (4.13.1) but with zero left-hand side instead of $\phi(\mathbf{R}_1)$, and integration by parts, as above, puts these relations into the form (4.13.3), but with zero left-hand sides. On adding these results for all the domains V_1, V_2, etc., since $\mathbf{v}.\boldsymbol{\nu}$ and the linearized pressure, $-\rho_0\mathbf{U}.\mathbf{v}$, are continuous everywhere except at the mean body surface, the surface integrals cancel everywhere except on Σ, and hence

$$\phi(\mathbf{R}_1) = -\frac{1}{4\pi}\int_{\Sigma} \frac{\mathbf{v}(\mathbf{R}).\boldsymbol{\nu}}{R_\beta}\,\mathrm{d}S - \frac{1}{4\pi}\int_{\Sigma} \mathbf{k}.\mathbf{v}(\mathbf{R})\frac{(\mathbf{R}-\mathbf{R}_1).\boldsymbol{\nu}}{(x-x_1)^2+(y-y_1)^2} \times\left(1+\frac{z_1-z}{R_\beta}\right)\mathrm{d}S - \frac{1}{4\pi}\int_{V}\frac{Q(\mathbf{R})}{R_\beta}\,\mathrm{d}V, \quad (4.13.4)$$

where the surface integrations are over *both sides* of Σ, $\boldsymbol{\nu}$ is the *outward* normal to Σ, and the volume integral is over all space.

If $\mathbf{v}_1$ and $\mathbf{v}_2$ are the perturbation velocities on side 1 and side 2 of Σ respectively,

$$\Delta\mathbf{v}=\mathbf{v}_2-\mathbf{v}_1, \tag{4.13.5}$$

and $\boldsymbol{\nu}$ is the normal directed from side 1 to side 2, then (4.13.4) can be written

$$\phi(\mathbf{R}_1)=-\frac{1}{4\pi}\int_\Sigma \frac{\Delta\mathbf{v}.\boldsymbol{\nu}}{R_\beta}\,\mathrm{d}S-\frac{1}{4\pi}\int_\Sigma \Delta\mathbf{v}.\mathbf{k}\frac{(\mathbf{R}-\mathbf{R}_1).\boldsymbol{\nu}}{(x-x_1)^2+(y-y_1)^2} \times\left(1+\frac{z_1-z}{R_\beta}\right)\mathrm{d}S-\frac{1}{4\pi}\int_V \frac{Q(\mathbf{R})}{R_\beta}\,\mathrm{d}V, \tag{4.13.6}$$

where the surface integrals are now over *one side* of Σ only. This is the required adaptation of the general solution of § 2.6. In this formula, $\Delta\mathbf{v}.\boldsymbol{\nu}$ and $\Delta\mathbf{v}.\mathbf{k}$ can be specified independently on Σ, and it can be verified that $\phi(\mathbf{R}_1)$ does give the correct discontinuities on Σ.

An analogous expression can be obtained for the potential in supersonic flow past a body whose mean surface is given. With surfaces S_1, S_2, etc., and domains V_1, V_2, etc., defined as above, if $\mathbf{R}_1$ is a point of V_1, say, then (3.6.12) gives

$$\phi(\mathbf{R}_1)=\frac{1}{2\pi}\int_{S_1'} \frac{\mathbf{w}(\mathbf{R}).\boldsymbol{\nu}_1}{R_B}\,\mathrm{d}S-\frac{B^2}{2\pi}{}^*\!\!\int_{S_1} \phi(\mathbf{R})\frac{(\mathbf{R}-\mathbf{R}_1).\boldsymbol{\nu}_1}{R_B^3}\,\mathrm{d}S -\frac{1}{2\pi}\int_{V_1'} \frac{Q(\mathbf{R})}{R_B}\,\mathrm{d}V, \tag{4.13.7}$$

where S_1' is that part of the bounding surface of V_1 which lies in the dependence domain of $\mathbf{R}_1$, V_1' is that part of V_1 which lies in the dependence domain of $\mathbf{R}_1$, and $\boldsymbol{\nu}_1$ is the outward normal to S_1'. The second integral can be integrated by parts with respect to z to give

$$\frac{1}{2\pi}\int_{S_1'} \frac{\partial\phi}{\partial z}\frac{(\mathbf{R}-\mathbf{R}_1).\boldsymbol{\nu}_1}{(x-x_1)^2+(y-y_1)^2}\frac{z_1-z}{R_B}\,\mathrm{d}S, \tag{4.13.8}$$

the line integrals vanishing due to the boundary condition (ix) of § 4.5, and on taking the finite part. Since $\boldsymbol{\nu}_1.\mathbf{w}=\boldsymbol{\nu}_1.\mathbf{v}$ on S_1', (4.13.7) becomes

$$\phi(\mathbf{R}_1)=\frac{1}{2\pi}\int_{S_1'} \frac{\mathbf{v}(\mathbf{R}).\boldsymbol{\nu}_1}{R_B}\,\mathrm{d}S +\frac{1}{2\pi}\int_{S_1'} \mathbf{k}.\mathbf{v}(\mathbf{R})\frac{(z_1-z)(\mathbf{R}-\mathbf{R}_1).\boldsymbol{\nu}}{[(x-x_1)^2+(y-y_1)^2]R_B}\,\mathrm{d}S-\frac{1}{2\pi}\int_{V_1'} \frac{Q(\mathbf{R})}{R_B}\,\mathrm{d}V. \tag{4.13.9}$$

The arguments leading to (4.13.6) can now be repeated to give the final result

$$\phi(\mathbf{R}_1) = -\frac{1}{2\pi}\int_\Sigma \frac{\Delta\mathbf{v}.\boldsymbol{\nu}}{R_B}\,\mathrm{d}S - \frac{1}{2\pi}\int_\Sigma \Delta\mathbf{v}.\mathbf{k}\,\frac{(z_1-z)(\mathbf{R}-\mathbf{R}_1).\boldsymbol{\nu}}{[(x-x_1)^2+(y-y_1)^2]\,R_B}\,\mathrm{d}S - \frac{1}{2\pi}\int_V \frac{Q(\mathbf{R})}{R_B}\,\mathrm{d}V, \quad (4.13.10)$$

where the surface integrations are over *one side* of that part of Σ which lies in the dependence domain of $\mathbf{R}_1$, and the volume integration is over the dependence domain of $\mathbf{R}_1$. As for the subsonic analogue, $\Delta\mathbf{v}.\boldsymbol{\nu}$ and $\Delta\mathbf{v}.\mathbf{k}$ can be specified independently on Σ, and it can be verified that $\phi(\mathbf{R}_1)$ does give the correct discontinuities on Σ.

The surface integrals in (4.13.6) and (4.13.10) can be interpreted physically. In both formulae, the first integral gives the potential field due to a distribution of sources on Σ, of surface density $\Delta\mathbf{v}.\boldsymbol{\nu}$. The second integral gives the potential field of a surface distribution of normally directed elementary vortex pairs on Σ, of surface density $\Delta\mathbf{v}.\mathbf{k}$. In most applications $Q=0$, and the volume integral does not occur.

When $\Delta\mathbf{v}.\mathbf{k}=0$ on Σ, then there are no vortex sheets in the flow, there is no pressure difference across Σ, and hence there is no lateral force on the body. A specification of the thickness distribution of the body on Σ then gives $\Delta\mathbf{v}.\boldsymbol{\nu}$ from the boundary condition (i) of § 4.5, and ϕ can be determined from (4.13.6) or (4.13.10); of course the body is not completely specified by giving the thickness distribution alone.

When $\Delta\mathbf{v}.\boldsymbol{\nu}=0$, then the body is of zero thickness; a specification of the normal force distribution on Σ gives $\Delta\mathbf{v}.\mathbf{k}$ from the linearized Bernoulli equation, and ϕ can be determined from (4.13.6) or (4.13.10). The geometry of the body then has to be determined from the calculated value of $\mathbf{v}.\boldsymbol{\nu}$ on Σ.

The above considerations show that (4.13.6) and (4.13.10) cannot be used to calculate the flow past a given body except in special cases. One such special case is that of plane symmetrical wings considered in §§ 5.6 and 6.4.

4.14. A variational principle for lifting surfaces

An interesting variational principle for lifting surfaces has been developed by Flax (1952*a*), which has particular applications to the approximate solutions of problems in subsonic flow past plane wings. As originally formulated by Flax, the principle applied to plane wings only, but the generalization to non-plane lifting surfaces is comparatively trivial.

The lifting surface problem deals with bodies of zero thickness for which the normal component of velocity at the surface is continuous through the surface, and the pressure is discontinuous. Let Σ be the mean surface for lifting surfaces of zero thickness in reciprocal flows. Then if p_1' and p_2' are the pressure differences across Σ in the two flows, the normal components of the perturbation velocities on Σ are linear functionals of the pressure differences in linearized theory, and the equations expressing this can be written symbolically in the form

$$\boldsymbol{\nu}.\mathbf{v}_1 = F_1\{p_1'\}, \quad \boldsymbol{\nu}.\mathbf{v}_2 = F_2\{p_2'\}. \tag{4.14.1}$$

For given surfaces (having the same mean surface) in the two flows, $\boldsymbol{\nu}.\mathbf{v}_1$ and $\boldsymbol{\nu}.\mathbf{v}_2$ are known on Σ, and (4.14.1) are integral equations for p_1' and p_2' whose solutions under the conditions that p_1' and p_2' vanish on subsonic trailing edges give the load distributions on the surfaces. The precise forms of the functionals in (4.14.1) are immaterial for the present purpose, but one form can be derived from the general solutions given above in §4.13; in fact, Flax (1952*a*) used other forms.

Now consider the integral

$$I = \int_\Sigma (p_1'\boldsymbol{\nu}.\mathbf{v}_2 + p_2'\boldsymbol{\nu}.\mathbf{v}_1 - p_2'F_1\{p_1'\})\,\mathrm{d}S, \tag{4.14.2}$$

where the integration is taken over one side of Σ. Then the variation of I, consequent upon giving p_1' and p_2' small variations $\delta p_1'$ and $\delta p_2'$ respectively, is

$$\delta I = \int_\Sigma (\delta p_1'\boldsymbol{\nu}.\mathbf{v}_2 + \delta p_2'\boldsymbol{\nu}.\mathbf{v}_1 - \delta p_2'F_1\{p_1'\} - p_2'F_1\{\delta p_1'\})\,\mathrm{d}S, \tag{4.14.3}$$

provided that the integral converges; a sufficient condition for this is that $\delta p_1'$ and $\delta p_2'$ should vanish at subsonic trailing edges in their respective flows.

For lifting surfaces, the flow-reversal theorem (4.10.4) can be written in the form

$$\int_\Sigma (p_1' F_2\{p_2'\} - p_2' F_1\{p_1'\}) \, dS = 0, \qquad (4.14.4)$$

where the integration is over *one side* of Σ. On varying p_1' in such a manner that the variation $\delta p_1'$ and its associated velocity field satisfies the Kutta-Joukowski condition at subsonic trailing edges, (4.14.4) gives

$$\int_\Sigma (\delta p_1' F_2\{p_2'\} - p_2' F_1\{\delta p_1'\}) \, dS = 0. \qquad (4.14.5)$$

The functional $F_1\{\delta p_1'\}$ can now be eliminated from (4.14.3) and (4.14.5) to give

$$\delta I = \int_\Sigma [\delta p_1'(\boldsymbol{\nu}.\mathbf{v}_2 - F_2\{p_2'\}) + \delta p_2'(\boldsymbol{\nu}.\mathbf{v}_1 - F_1\{p_1'\})] \, dS, \qquad (4.14.6)$$

or, by virtue of (4.14.1),

$$\delta I = \delta \int_\Sigma (p_1' \boldsymbol{\nu}.\mathbf{v}_2 + p_2' \boldsymbol{\nu}.\mathbf{v}_1 - p_2' F_1\{p_1'\}) \, dS = 0, \qquad (4.14.7)$$

which is Flax's form of the variational principle for lifting surfaces. It will be clear from the above derivation that other forms are possible.

PART II. SPECIAL METHODS

CHAPTER 5

SUBSONIC FLOW PAST THIN BODIES

5.1. Introduction

It is shown in § 2.3 that problems in the linearized theory of subsonic flow past thin bodies can be transformed to problems in incompressible flow past affinely distorted bodies; thus to give a general treatment of linearized subsonic flow problems would be to repeat in essentials the whole existing theory of incompressible flow past thin bodies. For this reason the subject-matter of this chapter mainly consists of applications of the transformation theory in a few typical cases, the exceptions being the calculation of induced drag from the vorticity in a wake in § 5.2, and the direct specification of the boundary-value problem for a nearly plane wing in § 5.6. (For subsonic flow past slender bodies, see Chapter 9.)

5.2. Vortex wakes and drag

It has been shown, in § 4.6, that the drag force on a body in subsonic flow is given by

$$D=\rho_0\int_S (\tfrac{1}{2}\mathbf{v}.\mathbf{wk}.\mathbf{n}-\mathbf{k}.\mathbf{vw}.\mathbf{n})\,\mathrm{d}S, \qquad (5.2.1)$$

where S is any closed surface surrounding the body, the integral being invariant for changes of S even when vortex sheets are present.

Now (5.2.1) is derived on the assumption that S is not too far away from the body, so that the linearized solution for $\mathbf{v}$ gives a uniform first approximation to the perturbation velocity on S (except perhaps at the edges of vortex sheets). However, since the integral is invariant for changes of S when $\mathbf{v}$ is the linear approximation, there is no reason why the drag should not be evaluated from the form taken by the *mathematical* solution at great distances from the body, even though $\mathbf{v}$ is no longer a good first approximation to the actual flow there. In this case, take the closed surface S to consist of $S_1+S_2+S_3$, where S_1, S_2, S_3 are the surfaces defined in § 4.9 in connexion with the uniqueness proof for subsonic flow; S is then a circular cylinder of large radius R_0 and great length

$2R_0$, and its plane ends. Then in the limit as $R_0 \to \infty$, by virtue of condition (viii) of §4.5, which gives the behaviour of $\mathbf{v}$ at great distances from the body, (5.2.1) becomes

$$D = \rho_0 \int_{S_3} (u^2 + v^2)\,dS = \rho_0 \int_{S_3} \left\{ \left(\frac{\partial \phi}{\partial x}\right)^2 + \left(\frac{\partial \phi}{\partial y}\right)^2 \right\} dS, \quad (5.2.2)$$

where S_3 is the plane at infinity downstream from the body, sometimes called the Trefftz plane.

If there are no vortex sheets in the flow, then $\partial\phi/\partial x$ and $\partial\phi/\partial y$ are both $O(1/R_0^3)$ as $R_0 \to \infty$, from (viii), §4.5, and hence, from (5.2.2),

$$D = 0. \quad (5.2.3)$$

If there are vortex sheets in the flow, let $\phi = \phi_1 + \phi_2$, where ϕ_1 is the potential due to any sources (i.e. the first integral in (4.13.6)) and ϕ_2 is the potential due to the elementary vortex pairs (i.e. the second integral in (4.13.6)). Then

$$\left(\frac{\partial \phi_1}{\partial x}\right)^2 = O(R_0^{-6}) \quad \text{and} \quad \frac{\partial \phi_1}{\partial x}\frac{\partial \phi_2}{\partial x} = O(R_0^{-3}) \quad (5.2.4)$$

on S_3 as $R_0 \to \infty$, and similarly for the y-derivatives. Therefore (5.2.2) becomes

$$D = \rho_0 \int_{S_3} \left\{ \left(\frac{\partial \phi_2}{\partial x}\right)^2 + \left(\frac{\partial \phi_2}{\partial y}\right)^2 \right\} dS, \quad (5.2.5)$$

the remaining terms vanishing by (5.2.4), and by the fact that (5.2.5) must converge (condition (vii) of §4.5).

Let the linearized vortex sheet intersect S_3 in a curve L, and let C be a contour described in the positive sense consisting of both sides of L and small circles of ultimately vanishing radius enclosing its ends, if any, where the velocity is infinite. Then, since $\partial\phi/\partial x$ and $\partial\phi/\partial y$ are both $O(1/r^2)$ (where $r^2 = x^2 + y^2$) as $r \to \infty$, an application of the two-dimensional form of Green's theorem to (5.2.5) gives

$$D = \rho_0 \int_C \phi \frac{\partial \phi}{\partial n}\,ds, \quad (5.2.6)$$

in which the suffix 2 has been dropped, as it is no longer required. The small circles contribute nothing to the integral in the limit as their radii vanish, so C in (5.2.6) is simply both sides of L. If L_+ and L_- denote the opposite sides of L, and with corresponding suffices for ϕ, (5.2.6) can be written

$$D = \rho_0 \int_{L_+} (\phi_+ - \phi_-) \frac{\partial \phi_+}{\partial n_+}\,ds, \quad (5.2.7)$$

since the normal component of velocity is continuous on L.

The velocity field in S_3 at infinity is that for a two-dimensional vortex distribution on L. With s as distance measured along L from some fixed point on L, let $\omega(s)$ be the vortex strength, or linear density of vorticity, on L. Then, if $\mathbf{r}'$ and $\mathbf{r}''$ are the position vectors in S_3 of two points s' and s'' respectively on L, the perturbation velocity $\mathbf{v}(s')$ at $\mathbf{r}'$ due to the element of vorticity $\omega(s'')\,ds''$ at $\mathbf{r}''$ is

$$\frac{\mathbf{k}\wedge(\mathbf{r}'-\mathbf{r}'')}{2\pi(\mathbf{r}'-\mathbf{r}'')^2}\,\omega(s'')\,ds''; \tag{5.2.8}$$

also

$$\mathbf{n}_+\,ds=\mathbf{k}\wedge\mathbf{ds}', \tag{5.2.9}$$

where $\mathbf{ds}'$ is the vector element of L at $\mathbf{r}'$. The equation (5.2.7) now gives, on using (5.2.8) and (5.2.9),

$$\begin{aligned} D&=\frac{\rho_0}{2\pi}\iint(\phi_+-\phi_-)\frac{\mathbf{k}\wedge(\mathbf{r}'-\mathbf{r}'').\mathbf{k}\wedge\mathbf{ds}'}{(\mathbf{r}'-\mathbf{r}'')^2}\,\omega(s'')\,ds'' \\ &=\frac{\rho_0}{2\pi}\iint(\phi_+-\phi_-)\,\omega(s'')\frac{(\mathbf{r}'-\mathbf{r}'').\mathbf{ds}'}{(\mathbf{r}'-\mathbf{r}'')^2}\,ds'', \end{aligned} \tag{5.2.10}$$

where $\phi_+-\phi_-$ is a function of s', and the integrations are along L_+. Since

$$\frac{\mathbf{r}'-\mathbf{r}''}{(\mathbf{r}'-\mathbf{r}'')^2}=\nabla'(\log|\mathbf{r}'-\mathbf{r}''|), \tag{5.2.11}$$

where ∇' is the two-dimensional gradient operator with respect to $\mathbf{r}'$, integration by parts with respect to s' in (5.2.10) gives

$$D=-\frac{\rho_0}{2\pi}\iint\frac{\partial}{\partial s'}(\phi_+-\phi_-)\,\omega(s'')\log|\mathbf{r}'-\mathbf{r}''|\,ds'\,ds''. \tag{5.2.12}$$

But (2.8.5), or (2.8.6), shows that

$$\frac{\partial}{\partial s'}(\phi_+-\phi_-)=\omega(s'); \tag{5.2.13}$$

therefore (5.2.12) gives the drag in terms of the strength of the vortex sheet as

$$D=\frac{\rho_0}{2\pi}\iint\log\frac{1}{|\mathbf{r}'-\mathbf{r}''|}\,\omega(s')\,\omega(s'')\,ds'\,ds'', \tag{5.2.14}$$

where both integrations are over L, taken in the same direction. This is a generalization of Prandtl's well-known formula for the induced drag of a nearly plane lifting surface in incompressible flow. Since the linearized vortex sheet is a cylindrical surface with generators parallel to $\mathbf{U}$, the double integral in (5.2.14) can be evaluated over any cross-section of the vortex sheet by a plane perpendicular to **U**. In this form, the expression for the drag is independent of β, but ω depends on β for a given body, so there is implicit dependence on β.

5.3. Two-dimensional flows, Glauert-Prandtl rule

For two-dimensional flows where there is no dependence upon x say, the potential equation becomes

$$\frac{\partial^2\phi}{\partial y^2}+\beta^2\frac{\partial^2\phi}{\partial z^2}=0, \tag{5.3.1}$$

which has the general solution

$$\phi=\Re f(z+i\beta y), \tag{5.3.2}$$

f being an arbitrary analytic function of the complex variable $z+i\beta y$. For a given problem the function f can be determined most easily from a related incompressible flow solution by using the transformations of §§2.1 and 2.3.

A very important problem is the calculation of the aerodynamic forces on thin monoplane wings of small camber, set at a small incidence to the stream. Wings designed for use at subsonic speeds generally have a rounded leading edge, in a neighbourhood of which the assumption of small surface slope is clearly violated. There is a stagnation point at which the perturbation velocity is $-\mathbf{U}$, so the assumption of small perturbations is also violated. However, the region of invalidity is small, and the errors in the calculated overall forces may also be expected to be small if the momentum method of §4.6 is used. That this is so can be confirmed by a study of more exact solutions (Lighthill, 1951).

In deriving the compressible flow past a two-dimensional wing from a corresponding incompressible flow, the choice of C and k can be made independently, because the method of §2.3 is valid. In order to show that the final result is the same, whatever values are given to C and k, they are not specified in the following derivations.

Consider a thin wing whose leading edge coincides with the x-axis, and which has a uniform section of chord c. Let this wing be set at an incidence α to the stream, the incidence being a right-handed rotation about the x-axis from the standard position for which the wing experiences no force in the direction of the y-axis. Let c', α', be the chord and incidence for a corresponding wing in an incompressible flow. Both wings lie approximately in the plane $y=0$, so the undisturbed streamlines which become the dividing streamlines lie in $y=0$.

With the notation of § 2.3, the distortion of the streamlines in the incompressible flow is $1/C$ times the distortion in the compressible flow, and the lengths in the stream direction are multiplied by a factor k/β.

The wing sections in the compressible and incompressible flows are shown (with the dimensions in the y-direction much exaggerated) in fig. 5.1. From this figure it is seen that

$$\alpha' = \frac{\alpha c}{C}\Big/\frac{kc}{\beta} = \frac{\beta}{Ck}\alpha, \qquad (5.3.3)$$

which relates the incidences in the two flows.

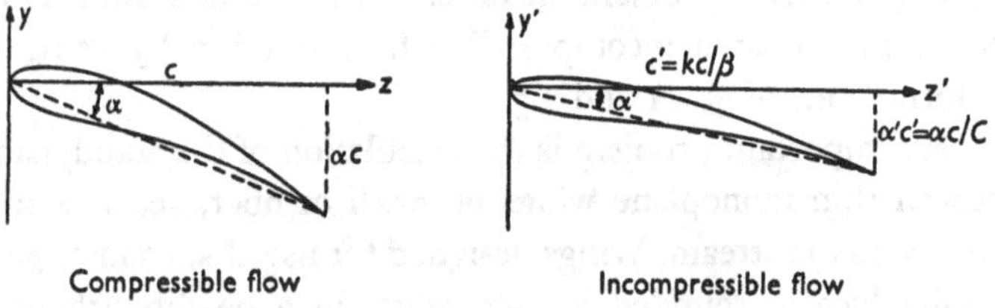

Fig. 5.1. Aerofoil sections in related two-dimensional compressible and incompressible flows.

The lift forces L, L', per unit length of wing measured in the x-direction, are

$$L = \oint (p-p_0)\,dz, \quad L' = \oint (p_i - p_0)\,dz', \qquad (5.3.4)$$

and the moments M, M', per unit length of wing, taken about the leading edges, are

$$M = \oint (p-p_0)\,z\,dz, \quad M' = \oint (p_i - p_0)\,z'\,dz', \qquad (5.3.5)$$

where the integrals are all taken round the section profiles. The drag forces, D, D', vanish for perfect fluids, since there are no vortex sheets in the flow.

The quantities most usually required are the non-dimensional lift and moment coefficients:

$$C_L = \frac{L}{\frac{1}{2}\rho_0 U^2 c} = -\frac{2}{cU}\oint \frac{\partial \phi}{\partial z}\,dz, \quad C_L' = \frac{L'}{\frac{1}{2}\rho_0 U^2 c'} = -\frac{2}{c'U}\oint \frac{\partial \phi_i}{\partial z'}\,dz'; \qquad (5.3.6)$$

$$\left.\begin{aligned} C_M &= \frac{M}{\frac{1}{2}\rho_0 U^2 c^2} = -\frac{2}{c^2 U}\oint \frac{\partial \phi}{\partial z} z\,\mathrm{d}z, \\ C'_M &= \frac{M'}{\frac{1}{2}\rho_0 U^2 c'^2} = -\frac{2}{c'^2 U}\oint \frac{\partial \phi_i}{\partial z'} z'\,\mathrm{d}z'. \end{aligned}\right\} \tag{5.3.7}$$

By transforming these integrals, for example,

$$-\frac{2}{cU}\oint \frac{\partial \phi}{\partial z}\,\mathrm{d}z = -\frac{2k}{\beta c' U}\oint \frac{Ck}{\beta^2}\frac{\partial \phi_i}{\partial z'}\frac{\beta}{k}\,\mathrm{d}z' = -\frac{Ck}{\beta^2}\frac{2}{c'U}\oint \frac{\partial \phi_i}{\partial z'}\,\mathrm{d}z', \tag{5.3.8}$$

it follows that

$$C_L = (Ck/\beta^2)\,C'_L, \quad C_M = (Ck/\beta^2)\,C'_M. \tag{5.3.9}$$

Therefore, by using (5.3.3),

$$\left(\frac{\partial C_L}{\partial \alpha}\right)_{\alpha=0} = \frac{1}{\beta}\left(\frac{\partial C'_L}{\partial \alpha'}\right)_{\alpha'=0}, \quad \left(\frac{\partial C_M}{\partial \alpha}\right)_{\alpha=0} = \frac{1}{\beta}\left(\frac{\partial C'_M}{\partial \alpha'}\right)_{\alpha'=0}. \tag{5.3.10}$$

Hence, for a given incidence, a thin two-dimensional wing in compressible flow has lift and moment coefficients which are $1/\beta$ times as great as those for a thin two-dimensional wing at the same incidence in incompressible flow. The thicknesses of the sections are different in general, but it is known that the lift and moment coefficients for a thin wing in two-dimensional incompressible flow are independent of the thickness of the section to a linear approximation, and this must be true for linearized compressible flow also.

The integrals for L and L' in (5.3.4) can be transformed in the usual way to give

$$L = CL'/\beta, \tag{5.3.11}$$

and it is shown in § 2.3 that, if K and K' are the circulations in the two flows, then

$$K = CK'/\beta. \tag{5.3.12}$$

Thus from the Kutta-Joukowski relation between lift and circulation for incompressible flow, namely,

$$L' = \rho_0 UK', \tag{5.3.13}$$

and from (5.3.11) and (5.3.12), it follows that

$$L = \rho_0 UK, \tag{5.3.14}$$

so the Kutta-Joukowski relation is true for linearized subsonic compressible flow.

If $C = 1$ and $k = \beta$, then there is no relative distortion of the streamlines, nor of lengths measured in the stream direction, so the wing sections are the same in both flows. Thus, for a given wing section, the effect of compressibility is to increase the surface perturbation pressures, the circulation, the lift, and the moment in incompressible flow by the factor $1/\beta = 1/\sqrt{(1-M^2)}$. This is the Glauert-Prandtl rule.†

On the Kutta-Joukowski hypothesis, the theoretical value of the lift coefficient for a thin wing section at incidence α in incompressible flow is $2\pi\alpha$, and the position of the centre of lift is at quarter-chord downstream from the leading edge. Hence the lift coefficient in linearized subsonic compressible flow is

$$C_L = \frac{2\pi\alpha}{\sqrt{(1-M^2)}}, \tag{5.3.15}$$

and the centre of lift remains at the quarter-chord point.

Clearly the Glauert-Prandtl rule cannot be expected to apply when M is too close to unity, since $1/\beta \to \infty$ as $M \to 1$. The range of applicability depends upon the thickness, camber and incidence of the wing section under consideration, but the rule can usually be relied upon up to about $M = 0{\cdot}7$ for moderate thickness, etc. (say about $0{\cdot}05$). The failure of linearized theory in this case, and indeed in all problems, for Mach numbers near to unity, is due to the non-uniform convergence at $M = 1$ of the series expansion for the potential in powers of the thickness, t, as in § 1.8. The problem of finding suitable alternative expansions which converge uniformly at $M = 1$ has not yet been solved. In any case, it appears that transonic flows, in which both subsonic and supersonic flow regions occur, present essentially non-linear problems, because the differential equations of motion must change from elliptic to hyperbolic type in the flow field.

5.4. Two-dimensional swept wings

If the infinite two-dimensional wing of § 5.3 is rotated about the y-axis through an angle χ, it is said to be swept at this angle. In order to determine the flow past such a swept wing, take new axes of rectangular cartesian co-ordinates x', y', z', such that the x'-axis lies along the leading edge, and the y'-axis coincides with the y-axis. Then

$$\left.\begin{aligned} x' &= x\cos\chi - z\sin\chi,\\ y' &= y,\\ z' &= z\cos\chi + x\sin\chi. \end{aligned}\right\} \tag{5.4.1}$$

† H. Glauert (1928), L. Prandtl (1930).

The perturbation velocity potential, ϕ, must be independent of the co-ordinate x', and the linearized equation for ϕ in the new co-ordinate system becomes

$$\frac{\partial^2\phi}{\partial y'^2}+(1-M^2\cos^2\chi)\frac{\partial^2\phi}{\partial z'^2}=0. \tag{5.4.2}$$

This equation has the general solution

$$\phi=\Re f(z'+i\beta'y'), \tag{5.4.3}$$

where $\beta'^2=1-M^2\cos^2\chi$. The pressure can be calculated from the linear approximation to Bernoulli's equation, which becomes

$$(p-p_0)/\rho_0=-\cos\chi U\,\partial\phi/\partial z'. \tag{5.4.4}$$

It is interesting to compare the pressures at a point on the wing for the unswept and swept positions. If 0, m, n are the direction cosines of the outward normal at any point on the unswept wing for the position of zero lift, then these are also the direction cosines relative to the x'-, y'-, z'-axes for the swept wing, and the direction cosines relative to the x-, y-, z-axes are $n\sin\chi, m, n\cos\chi$. The incidence, α, is measured from this standard position as a rotation about the x-axis in the positive sense, so the boundary condition on the mean wing surface is (cf. (1.9.7))

$$\partial\phi/\partial y'=-(n\cos\chi+\alpha)\,U, \tag{5.4.5}$$

since $m\simeq 1$ for a thin body. The effect of the incidence α can be separated from the effect of the thickness of the wing, and it follows from (5.4.3), (5.4.4) and (5.4.5), that for zero incidence the effect is to multiply the perturbation pressure at a given point on the unswept wing by a factor

$$\cos^2\chi\sqrt{\left(\frac{1-M^2}{1-M^2\cos^2\chi}\right)}, \tag{5.4.6}$$

which is always less than unity. This factor is independent of the wing shape, so there is no force on the swept wing in this standard position. The extra pressure due to incidence at a given point on the swept wing is that at the same point on the unswept wing multiplied by the factor

$$\cos\chi\sqrt{\left(\frac{1-M^2}{1-M^2\cos^2\chi}\right)}, \tag{5.4.7}$$

which is also less than unity. Hence the lift-coefficient for the swept wing on the Kutta-Joukowski hypothesis is

$$C_L = \frac{2\pi\alpha\cos\chi}{\sqrt{(1-M^2\cos^2\chi)}}. \tag{5.4.8}$$

If the two wings are not the same in the swept and unswept positions, but have the same sections by any plane $x = constant$, then the factor between the perturbation pressures is always (5.4.7) both for zero incidence and incidence α. The lift coefficient is still given by (5.4.8).

5.5. The lifting line theory for wings of finite span

Just as for incompressible flow, unswept monoplane wings of finite span in subsonic compressible flow can be treated by Prandtl's lifting line theory. The results for compressible flow can be obtained directly from those for incompressible flow by means of the transformation theory of § 2.3. In making the transformation, it is convenient to keep the circulations and the span the same, so that $C=\beta$ and $k=1$. The pressure is then increased from the incompressible value by a factor $1/\beta$, and the chord is decreased by a factor β; hence the lift per unit length in the spanwise direction remains unaltered, as is to be expected from the constancy of the circulation. The area of the planform is decreased by a factor β, so the aspect ratio is increased by a factor $1/\beta$. This last fact limits the range of applicability of the theory for compressible flow because the aspect ratio in the incompressible flow is β times less than the given aspect ratio in compressible flow. Prandtl's theory assumes a very large aspect ratio, but it is found experimentally that the theory gives quite accurate results for aspect ratios as small as 4 in incompressible flow. Hence it may be inferred that the corresponding lower limit in compressible flow is about $4/\beta$. If $M = 0{\cdot}7$, then $4/\beta \simeq 5{\cdot}6$, which is much the same as the aspect ratio of some modern high-speed wing planforms.

The lifting line theory for incompressible flow is developed in detail in standard text-books.† The only correction required is to the fundamental formula

$$K' = \tfrac{1}{2}a_0'c'(U\alpha + v'), \tag{5.5.1}$$

† See, for example, H. Glauert, *The Elements of Aerofoil and Airscrew Theory* (Cambridge, 1937), or L. M. Milne-Thomson, *Theoretical Aerodynamics* (Macmillan, 1948).

which applies at all sections; K' is the circulation, a_0' is $(\partial C_L'/\partial\alpha')_{\alpha'=0}$ for the section, c' is the chord of the section, α' is the geometrical angle of incidence of the section, and $-v'$ is the downwash at the lifting line. Equation (5.5.1) is an expression of theKutta-Joukowski relation (5.3.11) for a section of the wing. On transforming to compressible flow, with $C=\beta$, $k=1$,

$$K=K', \quad a_0=(\partial C_L/\partial\alpha)_{\alpha=0}=a_0'/\beta, \quad c=\beta c', \quad \alpha=\alpha', \quad v=v', \tag{5.5.2}$$

and the formula (5.5.1) becomes

$$K=\tfrac{1}{2}a_0c(U\alpha+v). \tag{5.5.3}$$

This is of the same form as before, but a_0 is now to be taken for compressible flow at the appropriate Mach number. From (5.3.13) the theoretical value of a_0 is

$$\frac{2\pi\alpha}{\sqrt{(1-M^2)}}, \tag{5.5.4}$$

but, as in incompressible flow, an experimentally determined value can be substituted instead. The downwash at the lifting line is given in terms of the circulation distribution along the span by the same formula as in incompressible flow theory, and the calculations follow the same lines.

A more detailed discussion of the problem, including an investigation of the flow conditions in the wake, has been given by Goldstein and Young (1943).

5.6. Nearly plane wings

In some cases, lifting line theory is not sufficiently accurate, or does not given sufficiently detailed information, in which cases it becomes necessary to solve a lifting surface problem that can be specified as follows.

Consider a thin wing whose mean surface, Σ, can be taken to be a plane; for convenience, this plane is supposed to be $y=0$ in the usual system of rectangular cartesian co-ordinates. Let l_+, m_+, n_+ and l_-, m_-, n_- be the direction cosines of the outward normals to the actual wing surfaces corresponding to the sides $y=+0$ and $y=-0$ of the mean surface. Then the boundary conditions of zero relative normal velocity on the wing surfaces are, from the results of § 1.9,

$$\left(\frac{\partial\phi}{\partial y}\right)_{y=+0}=-Un_+, \quad \left(\frac{\partial\phi}{\partial y}\right)_{y=-0}=Un_-. \tag{5.6.1}$$

Now let $\phi=\phi_1+\phi_2$, where ϕ_1 is an even function of y, and ϕ_2 is an odd function of y (ϕ_1 and ϕ_2 can always be chosen in this way). Then (5.6.1) gives

$$\left(\frac{\partial\phi_1}{\partial y}\right)_{y=+0}=-\left(\frac{\partial\phi_1}{\partial y}\right)_{y=-0}=-\tfrac{1}{2}U(n_++n_-), \tag{5.6.2}$$

$$\left(\frac{\partial\phi_2}{\partial y}\right)_{y=+0}=\left(\frac{\partial\phi_2}{\partial y}\right)_{y=-0}=-\tfrac{1}{2}U(n_+-n_-). \tag{5.6.3}$$

Thus the general problem is divided into two parts: one is to determine a symmetrical flow field, and the other is to determine an anti-symmetrical flow field.

The former, symmetrical, problem can be solved by using the results of §4.13. Since ϕ_1 is an even function of y, so are $\partial\phi_1/\partial x$ and $\partial\phi_1/\partial z$, while $\partial\phi_1/\partial y$ is an odd function of y; also the tangential components of velocity are continuous through Σ, and the normal component has a discontinuity $2(\partial\phi_1/\partial y)_{y=+0}$. Therefore, if there are no sources in the flow (i.e. $Q=0$), the formula (4.13.6) gives

$$\phi_1(\mathbf{R}_1)=-\frac{1}{2\pi}\iint_\Sigma\left(\frac{\partial\phi_1}{\partial y}\right)_{y=+0}\frac{dz\,dx}{\sqrt{\{\beta^2(x-x_1)^2+\beta^2y_1^2+(z-z_1)^2\}}}, \tag{5.6.4}$$

which determines ϕ_1 everywhere from a knowledge of n_+ and n_- on the wing surface. The velocity components can then be calculated by differentiation, either before or after integration. Usually only the pressure on the wing surface is required, which can be calculated on the mean surface, without causing further error, by using the linearized approximation to Bernoulli's equation. This part of the general problem then can be considered to be solved completely. It must be mentioned, however, that most subsonic wings have rounded leading edges, in neighbourhoods of which the assumptions of linearized theory are violated; however, outside these neighbourhoods, the solution can generally be relied upon.

The anti-symmetrical problem is much more difficult, and no solution of the above form has been given, although there are approximate numerical methods for finding the pressure distribution. Since ϕ_2 is an odd function of y, it follows that $\partial\phi_2/\partial x$ and $\partial\phi_2/\partial z$ are also odd functions of y, and that $\partial\phi_2/\partial y$ is an even function of y. The whole plane $y=0$ can be divided into three parts: (i) the mean wing surface, Σ; (ii) the linearized vortex sheet,

T, stretching downstream from the trailing edge of Σ; and (iii) the remaining area, R. On R and T, $\partial\phi_2/\partial z$ must be continuous and therefore vanishes there, being an odd function of y; also ϕ_2 is continuous on R and vanishes there for the same reason. Thus the boundary conditions on $y=0$ which are satisfied by ϕ_2 are

(i) $\partial\phi_2/\partial y$ takes given values on Σ, from (5.6.3),
(ii) $\phi_2=0$ on R,
(iii) $\partial\phi_2/\partial z=0$ on T,
(iv) $\partial\phi_2/\partial y$ is finite at the trailing edge of Σ (Kutta-Joukowski condition),

giving what is called a mixed boundary-value problem. Its direct analytical solution requires a knowledge of an appropriate Green's function, the determination of which is just as difficult as the solution of the boundary-value problem itself, and it would seem that this approach is not a very profitable one in general. Various numerical methods of solving the problem have been proposed, which enable the surface pressures to be determined to any required order of accuracy, but these are outside the scope of this monograph.

5.7. Quasi-steady flows; transformation of stability derivatives

Unsteady flows in which the time rates of change of the flow variables are sufficiently small to be neglected in the equations of motion are called quasi-steady flows. For a thin body that is moving slowly in a uniform steady stream, thus giving a quasi-steady flow, it is usual to calculate the flow from the linearized equations for steady flow by applying the fully linearized boundary conditions (modified appropriately on the mean body surface to include the effects of the small unsteady motion), which is equivalent to calculating the steady flow past a distorted body. Of course this distorted body must satisfy the usual conditions of thinness if linearized theory is to be applicable. Clearly the resulting solution of the linearized equations does not represent a uniform first approximation to the true flow as the distance from the body increases, because the finite speed of propagation of disturbances is not taken into account properly; also, in general, any linearized trailing vortex sheet does not approximate uniformly even to the ideal trailing vortex sheet as the distance from the body increases. The accuracy of the quasi-steady solution can be investigated in any particular case by expanding the velocity potential as a series in terms of some suitable parameter involving a measure of the small unsteadiness; but it can be seen on physical grounds that the solution gives the correct flow on and near the body, within the linearized order of approximation, provided that the time lags due to the finite speed of sound would cause only small proportional changes in the perturbation velocities at the surface of the body, and, in practice, criteria for the maximum allowable unsteadiness can be derived from this consideration.

The theory of quasi-steady flows can be used in conjunction with the transformation theory of Chapter 2 to calculate stability derivatives for thin bodies in subsonic flows from the corresponding quantities for affinely distorted bodies in incompressible flows. As an example, consider a nearly plane wing executing a steady roll with small angular velocity Ω. If s is the semi-span of the wing, then the greatest possible time taken for a disturbance to cross the wing is of order $s/(c_0-U)$, and in this time the velocity on the surface must not change significantly, which is equivalent to saying that the body must not change its position significantly in this case. The change of position of the wing tips is $\Omega s^2/(c_0-U)$, and this must be small compared with s, so a conservative criterion of smallness for Ω is that $\Omega s/U \ll (1-M)/M$. If the mean wing surface is in the plane $y=0$ at the time under consideration, and the axis of roll is the z-axis, then the boundary condition at the mean body surface is

$$\frac{\partial \phi}{\partial y} = -nU + \Omega x. \tag{5.7.1}$$

On transforming to incompressible flow by using (2.1.2), (2.2.3) and (2.2.4), and keeping the same rate of roll, this becomes

$$\frac{Ck}{\beta}\frac{\partial \phi_i}{\partial y'} = -\frac{n'U}{\beta} + \frac{\Omega x'}{k}, \tag{5.7.2}$$

which agrees with the corresponding boundary condition in incompressible flow if and only if $Ck=1$ and $Ck^2=\beta$, that is $C=1/\beta$, $k=\beta$.†

† Various alternative procedures are possible, for the two parameters C and k can be chosen at will, but this choice, following the choice of the same rate of roll for the transformed problem, ensures that the transformation is correct for both thin and slender bodies, and leads to the simplest calculation of the stability derivative.

The non-dimensional coefficient of rolling moment about the z-axis per unit rate of roll, usually denoted by l_p, is

$$l_p = \frac{1}{\rho_0 U\Omega s^2 S}\int (p-p_0)\, x \, dz\, dx, \tag{5.7.3}$$

where S is the wing area, and the integration is over the wing surface. On transforming to incompressible flow variables, since s transforms like x, and S transforms like $dz\,dx$, (5.7.3) gives

$$l_p = (1/\beta)\, l'_{p'}, \tag{5.7.4}$$

where $l'_{p'}$ is the corresponding derivative for the transformed wing in incompressible flow. If $l'_{p'}$ is a function of, say, aspect ratio, A', and sweep, χ', given by

$$l'_{p'} = F(A', \tan\chi'), \tag{5.7.5}$$

then, since $A' = A\beta$, and $\beta\tan\chi' = \tan\chi$, (5.7.4) and (5.7.5) give l_p as

$$l_p = (1/\beta)\, F(A\beta, \tan\chi/\beta). \tag{5.7.6}$$

CHAPTER 6

SUPERSONIC FLOW PAST NEARLY PLANE WINGS

6.1. Two-dimensional flows

In the absence of sources, the linearized equation for two-dimensional supersonic flow in the (y, z)-plane is

$$\frac{\partial^2\phi}{\partial y^2} - B^2\frac{\partial^2\phi}{\partial z^2} = 0. \tag{6.1.1}$$

If $G(y, z) = 0$ is the equation of the characteristic lines, then

$$\left(\frac{\partial G}{\partial y}\right)^2 - B^2\left(\frac{\partial G}{\partial z}\right)^2 = 0, \tag{6.1.2}$$

so the characteristic lines are straight lines of slope $\pm 1/B = \pm \tan\mu$, making an angle μ, the Mach angle, with the direction of the undisturbed stream. The general solution of (6.1.1) is

$$\phi = f(z - By) + g(z + By), \tag{6.1.3}$$

where f and g are arbitrary functions. Thus each term in (6.1.3) is constant along characteristic lines of one family.

Now consider a thin sharp-edged wing section of chord c, as shown in fig. 6.1, and take the origin of co-ordinates at the leading edge. OQ, OQ' and PR, PR' are characteristic or Mach lines from the leading and trailing edges. The disturbance caused by the wing must vanish upstream from QOQ' by the boundary condition (ix) of §4.5; hence, in $y > 0$, the function g in (6.1.3) must be identically zero; and similarly, in $y < 0$, the function f must vanish identically, otherwise this condition cannot be satisfied. In the region $QOPR$, the function f is determined by the boundary condition on the upper wing surface, and g is determined in $Q'OPR'$ by the boundary condition on the lower surface.

For an arbitrary attitude of the wing, let $\chi_+(z)$, $\chi_-(z)$ be the slopes of the upper and lower surfaces respectively, and if generally suffices + and − denote quantities in $y > 0$ and $y < 0$ respectively, then the linearized boundary conditions on the mean wing surface are

$$\left(\frac{\partial\phi_+}{\partial y}\right)_{y=+0} = -Bf'(z) = \chi_+(z)\,U \quad (0 < z < c), \tag{6.1.4}$$

$$\left(\frac{\partial \phi_-}{\partial y}\right)_{y=-0} = Bg'(z) = \chi_-(z)\,U \quad (0<z<c), \tag{6.1.5}$$

where dashes denote differentiation with respect to the argument of the functions. Therefore, in the region $QOPR$, $0<z-By<c$,

$$\frac{\partial \phi_+}{\partial y} = U\chi_+(z-By), \quad \frac{\partial \phi_+}{\partial z} = -\frac{U}{B}\chi_+(z-By), \tag{6.1.6}$$

and in the region $Q'OPR'$, $0<z+By<c$,

$$\frac{\partial \phi_-}{\partial y} = U\chi_-(z+By), \quad \frac{\partial \phi_-}{\partial z} = \frac{U}{B}\chi_-(z+By). \tag{6.1.7}$$

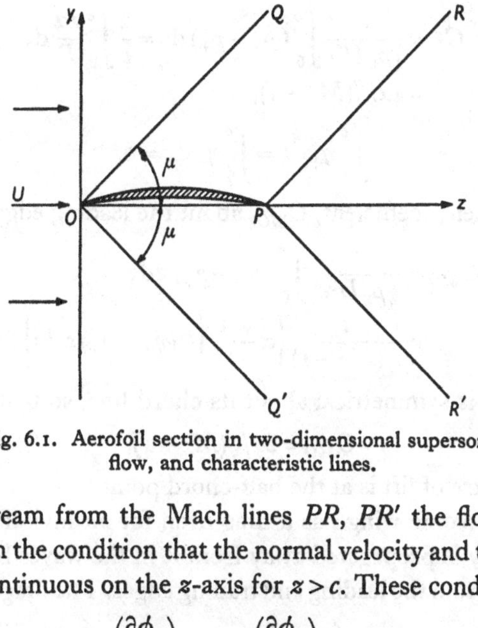

Fig. 6.1. Aerofoil section in two-dimensional supersonic flow, and characteristic lines.

Downstream from the Mach lines PR, PR' the flow is determined from the condition that the normal velocity and the pressure must be continuous on the z-axis for $z>c$. These conditions are

$$\left(\frac{\partial \phi_+}{\partial y}\right)_{y=+0} = \left(\frac{\partial \phi_-}{\partial y}\right)_{y=-0}, \tag{6.1.8}$$

and

$$\left(\frac{\partial \phi_+}{\partial z}\right)_{y=+0} = \left(\frac{\partial \phi_-}{\partial z}\right)_{y=-0}, \tag{6.1.9}$$

which lead to the result that ϕ_+ and ϕ_- are constants downstream from RPR', and hence the disturbance velocities are zero. If there is a circulation round the aerofoil, these constants are not equal, and the potential is multi-valued. There is no vortex sheet in linearized theory.

It is not difficult to show from (6.1.6) and (6.1.7) that the perturbation velocity is normal to the Mach lines.

If η_+ and η_- are the slopes of the upper and lower surfaces relative to the chord line OP, and α is the incidence of the chord line to the main stream, then

$$\chi_+ = \eta_+ - \alpha, \quad \chi_- = \eta_- - \alpha, \tag{6.1.10}$$

and the pressures on the upper and lower surfaces are respectively

$$\left.\begin{aligned} p_+ - p_0 &= \rho_0 U^2(\eta_+ - \alpha)/B, \\ p_- - p_0 &= -\rho_0 U^2(\eta_- - \alpha)/B. \end{aligned}\right\} \tag{6.1.11}$$

The lift coefficient, C_L, with respect to area, is

$$\begin{aligned} C_L &= \frac{1}{\frac{1}{2}\rho_0 U^2 c}\int_0^c (p_- - p_+)\,dz = \frac{2}{c}\int_0^c \frac{2\alpha}{B}\,dz \\ &= 4\alpha/\surd(M^2-1), \end{aligned} \tag{6.1.12}$$

since

$$\int_0^c \eta_+\,dz = \int_0^c \eta_-\,dz = 0. \tag{6.1.13}$$

The moment coefficient, C_{MO}, about the leading edge, is

$$\begin{aligned} C_{MO} &= \frac{1}{\frac{1}{2}\rho_0 U^2 c^2}\int_0^c (p_- - p_+)\,z\,dz \\ &= \frac{2}{\surd(M^2-1)}\left\{\alpha - \frac{1}{c^2}\int_0^c (\eta_+ + \eta_-)\,z\,dz\right\}. \end{aligned} \tag{6.1.14}$$

If the wing is symmetrical about its chord line, so that $\eta_+ = -\eta_-$, then

$$C_{MO} = 2\alpha/\surd(M^2-1), \tag{6.1.15}$$

and the centre of lift is at the half-chord point.

The supersonic wing has a drag even for an inviscid fluid, because energy is propagated away from it in the waves between the Mach lines from the leading and trailing edges. The drag coefficient, C_D, is

$$\begin{aligned} C_D &= \frac{1}{\frac{1}{2}\rho_0 U^2 c}\int_0^c (p_+\chi_+ + p_-\chi_-)\,dz \\ &= \frac{2}{\surd(M^2-1)}(2\alpha^2 + \overline{\eta_+^2} + \overline{\eta_-^2}), \end{aligned} \tag{6.1.16}$$

where the bars denote mean values over the chord.

The above results for lift, moment and drag coefficients in linearized theory were first obtained by Ackeret (1925). The formulae are not accurate near $M=1$, and when M is large. Near $M=1$, the expansion in a power series in the thickness parameter,

t, breaks down due to non-uniform convergence at $M=1$, as has been mentioned already (cf. §§ 1.8 and 5.3). The inaccuracy for large M is caused by the failure of the assumption of constant entropy through the discontinuity of velocity at the Mach lines OQ, OQ'; the entropy change is proportional to the cube of the strength of the discontinuity when this is small, but the factor of proportionality increases with M.

In the above arguments it is assumed that the leading edge of the wing is sharp. The velocity does not then become infinite at the leading edge in supersonic flow, as it does in subsonic flow, so this sharpness causes no difficulty. In practice the leading edge cannot be exactly sharp, and more exact considerations show that the shock is then much stronger near the leading edge than it is elsewhere; in the case of a very blunt or rounded leading edge, the bow shock is detached from the leading edge. As a result, the entropy changes through those portions of the shock near to the leading edge may be comparatively large, and these entropy variations are maintained along the streamlines in the vicinity of the wing surfaces, forming an 'entropy boundary layer'. The flow in this boundary layer is rotational, and the vorticity must be parallel to the leading edge in two-dimensional flow. Thus the component of vorticity in the stream direction is zero, and the results of § 1.12 apply; it follows from these results that the *pressure* on the wing surface is not altered outside the immediate neighbourhood of the leading edge, although the perturbation velocity may be altered considerably. Thus any slight bluntness of the leading edge is likely to have very little effect on force coefficients obtained by assuming the edge to be sharp (Lighthill, 1949*a*).

6.2. Two-dimensional swept wings

The effect of sweep on a two-dimensional wing in supersonic flow can be calculated in the same way as for the subsonic case (§ 5.4). If χ is the angle of sweep, then in the x', y', z' co-ordinate system, defined in (5.4.1), the equation for ϕ is (5.4.2), that is,

$$\frac{\partial^2\phi}{\partial y'^2}+(1-M^2\cos^2\chi)\frac{\partial^2\phi}{\partial z'^2}=0. \qquad (6.2.1)$$

Two cases arise, depending on whether $M^2\cos^2\chi-1$ is greater than or less than zero. If $M^2\cos^2\chi - 1 = B'^2 > 0$, then the relative motion is supersonic, and the lift coefficient taken with respect to area is

$$C_L = \frac{4\alpha\cos\chi}{\sqrt{(M^2\cos^2\chi - 1)}}, \tag{6.2.2}$$

the incidence α being a rotation about the x-axis, as usual. If the wing is the same as that of § 6.1, simply rotated through an angle χ about the y-axis, then the drag coefficient is

$$C_D = \frac{2\cos\chi}{\sqrt{(M^2\cos^2\chi - 1)}}\{2\alpha^2 + \cos^2\chi(\overline{\eta_+^2} + \overline{\eta_-^2})\}. \tag{6.2.3}$$

There is also a lateral force directed parallel to the x-axis, whose force coefficient, C_x, is given by

$$C_x = \frac{2\sin\chi\cos^2\chi}{\sqrt{(M^2\cos^2\chi - 1)}}(\overline{\eta_+^2} + \overline{\eta_-^2}). \tag{6.2.4}$$

Alternatively, if $1 - M^2\cos^2\chi = \beta'^2 > 0$, then the relative motion is subsonic, and the results of § 5.4 apply. There is no drag and the lift coefficient on the Kutta-Joukowski hypothesis is

$$C_L = \frac{2\pi\alpha\cos\chi}{\sqrt{(1 - M^2\cos^2\chi)}}. \tag{6.2.5}$$

It is easily shown from (6.2.2) and (6.2.3) that sweeping the wing improves the lift/drag ratio by a factor $\sec^2\chi$ for very small incidences, and improves the maximum value of lift/drag by a factor $\sec\chi$. The beneficial effect of sweep for supersonic wings was first noticed by Busemann (1935).

6.3. Nearly plane wings of finite span: the solutions for the potential

For problems of flow past nearly plane wings, the boundary data is given on the plane mean surface of the wing. The formula (3.6.12) for the potential can be adapted for this case so that only the potential, or its normal derivative, on the mean plane occurs in the expression for the potential at points not on the mean plane.

For convenience, suppose that the potential vanishes everywhere in $z \leqslant 0$, and that the boundary data are given on the plane $y = 0$, $z > 0$. Then for a point $\mathbf{R}_1 = [x_1, y_1, z_1]$ such that $y_1 > 0$, $z_1 > By_1$, (3.6.12) applied to V_1, the interior of the dependence domain of

$\mathbf{R}_1$ in $y>0$, bounded by the plane $z=0$, and the plane $y=0$ (see fig. 6.2), gives

$$\phi(\mathbf{R}_1) = -\frac{1}{2\pi}\int_{V_1}\frac{Q(\mathbf{R})}{R_B}\,\mathrm{d}V$$
$$-\frac{1}{2\pi}\iint_{\Sigma}\left(\frac{\partial\phi}{\partial y}\right)_{y=+0}\frac{\mathrm{d}z\,\mathrm{d}x}{\sqrt{\{(z-z_1)^2-B^2(x-x_1)^2-B^2y_1^2\}}}$$
$$-\frac{B^2}{2\pi}{}^*\!\!\iint_{\Sigma}\phi(x,+0,z)\frac{y_1\,\mathrm{d}z\,\mathrm{d}x}{\{(z-z_1)^2-B^2(x-x_1)^2-B^2y_1^2\}^{\frac{3}{2}}}, \tag{6.3.1}$$

where Σ is that part of $y=0$, $z>0$, which lies in the dependence domain of $\mathbf{R}_1$, and the surface integral over $z=0$ vanishes.

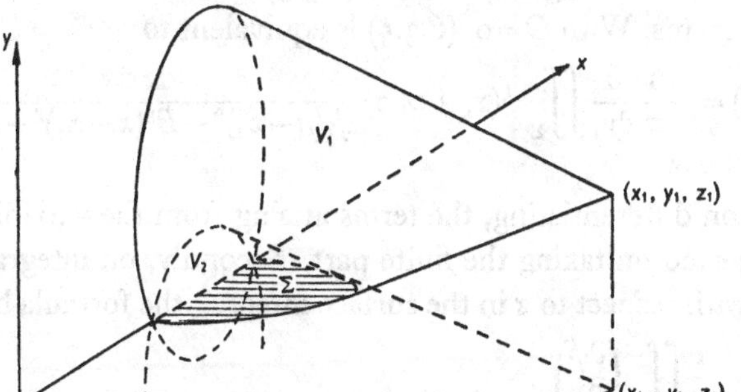

Fig. 6.2. Surfaces and volumes of integration in equations (6.3.4) and (6.3.5).

Now if $\mathbf{R}_1'$ is taken to be the point $(x_1, -y_1, z_1)$, and the domain of integration is V_2, the interior of the dependence domain of $\mathbf{R}_1'$ in $y>0$, bounded by the same two planes as before, then, since $\mathbf{R}_1'$ is an exterior point to V_2, (3.6.12) gives

$$0 = -\frac{1}{2\pi}\int_{V_2}\frac{Q(\mathbf{R})}{R_B'}\,\mathrm{d}V$$
$$-\frac{1}{2\pi}\iint_{\Sigma}\left(\frac{\partial\phi}{\partial y}\right)_{y=+0}\frac{\mathrm{d}z\,\mathrm{d}x}{\sqrt{\{(z-z_1)^2-B^2(x-x_1)^2-B^2y_1^2\}}}$$
$$+\frac{B^2}{2\pi}{}^*\!\!\iint_{\Sigma}\phi(x,+0,z)\frac{y_1\,\mathrm{d}z\,\mathrm{d}x}{\{(z-z_1)^2-B^2(x-x_1)^2-B^2y_1^2\}^{\frac{3}{2}}}, \tag{6.3.2}$$

where $$R_B' = \sqrt{\{(z-z_1)^2-B^2(x-x_1)^2-B^2(y+y_1)^2\}}. \tag{6.3.3}$$

By adding and subtracting (6.3.1) and (6.3.2), expressions are obtained for $\phi(\mathbf{R}_1)$ in terms of $(\partial\phi/\partial y)_{y=+0}$ and $\phi(x, +0, z)$; these are

$$\phi(\mathbf{R}_1) = -\frac{1}{\pi}\iint_\Sigma \left(\frac{\partial\phi}{\partial y}\right)_{y=+0} \frac{dz\,dx}{\sqrt{\{(z-z_1)^2 - B^2(x-x_1)^2 - B^2y_1^2\}}} - \frac{1}{2\pi}\int_{V_1} \frac{Q(\mathbf{R})}{R_B}\,dV - \frac{1}{2\pi}\int_{V_2} \frac{Q(\mathbf{R})}{R'_B}\,dV, \quad (6.3.4)$$

and
$$\phi(\mathbf{R}_1) = -\frac{B^2}{\pi}{}^*\!\!\iint_\Sigma \phi(x, +0, z) \frac{y_1\,dz\,dx}{\{(z-z_1)^2 - B^2(x-x_1)^2 - B^2y_1^2\}^{\frac{3}{2}}} - \frac{1}{2\pi}\int_{V_1} \frac{Q(\mathbf{R})}{R_B}\,dV + \frac{1}{2\pi}\int_{V_2} \frac{Q(\mathbf{R})}{R'_B}\,dV. \quad (6.3.5)$$

Usually there are no sources, and the volume integrals do not occur. Similar formulae apply to the half-space $y<0$.

The surface integral over Σ in (6.3.5) can be expressed in alternative forms. With $Q=0$, (6.3.5) is equivalent to

$$\phi(\mathbf{R}_1) = -\frac{1}{\pi}\frac{\partial}{\partial y_1}\iint_\Sigma \phi(x, +0, z) \frac{dz\,dx}{\sqrt{\{(z-z_1)^2 - B^2(x-x_1)^2 - B^2y_1^2\}}}, \quad (6.3.6)$$

since, on differentiating, the terms arising from the variable limits are rejected on taking the finite part. Secondly, on integrating by parts with respect to z in the surface integral, the formula becomes

$$\phi(\mathbf{R}_1) = \frac{1}{\pi}\iint_\Sigma \left(\frac{\partial\phi}{\partial z}\right)_{y=+0} \times \frac{y_1(z_1-z)\,dz\,dx}{\{(x-x_1)^2+(y-y_1)^2\}\sqrt{\{(z-z_1)^2 - B^2(x-x_1)^2 - B^2y_1^2\}}}, \quad (6.3.7)$$

since the line integral vanishes on taking the finite part (cf. (4.13.8)).

These results can also be obtained from (4.13.10).

For a nearly plane wing, whose mean surface lies in the plane $y=0$, the boundary conditions on this plane are identical with those given in §5.6 for subsonic flow, and the general problem can be divided into a symmetrical and an anti-symmetrical problem. These two problems can be treated separately, and the boundary conditions on the wing are given by (5.6.2) and (5.6.3) respectively.

6.4. The symmetrical problem

If ϕ_1 is the perturbation potential for flow past a given wing symmetrical about its chord plane, $y=0$, then ϕ_1 is an even function of y, and $\partial\phi_1/\partial y$ is an odd function of y. Since $\partial\phi_1/\partial y$ must be continuous everywhere except through the mean wing surface (i.e. the chord plane), the boundary conditions for ϕ_1 in $y>0$, say, are

(i) $\partial\phi_1/\partial y$ takes given values on the mean surface Σ for $y=+0$,

(ii) $\partial\phi_1/\partial y=0$ on the rest of $y=+0$.

The potential ϕ_1 is then given from (6.3.4) (with no sources, $Q=0$) as

$$\phi_1(x_1,y_1,z_1)=-\frac{1}{\pi}\iint_{\Sigma'}\left(\frac{\partial\phi_1}{\partial y}\right)_{y=+0}\frac{dz\,dx}{\sqrt{\{(z-z_1)^2-B^2(x-x_1)^2-B^2y_1^2\}}}, \tag{6.4.1}$$

where Σ' is that part of the mean wing surface for which

$$z\leqslant z_1-B\sqrt{\{(x-x_1)^2+y_1^2\}}. \tag{6.4.2}$$

The potential in $y_1<0$ is also determined, since ϕ_1 is an even function of y_1.

The result (6.4.1) for the potential in the symmetrical problem was given by Puckett (1946), who made a direct verification that this is the correct potential, and applied the formula to calculate the pressure distribution and drag force on some special wings. Given the result (6.4.1) the complete solution of the problem is straightforward mathematically, but the labour required is great. If only the drag is required for a given wing, the flow reversal result of §4.11 (i) sometimes reduces the amount of computation.

6.5. The anti-symmetrical problem: characteristic coordinates

As in §5.6, let T denote the linearized trailing vortex sheet in the plane $y=0$, and let R denote the remainder of the plane $y=0$. Then, since the perturbation potential ϕ_2 for the anti-symmetrical problem is an odd function of y, the boundary conditions on $y=+0$, say, for the potential ϕ_2 in $y>0$ are (cf. §5.6)

(i) $\partial\phi_2/\partial y$ takes given values on Σ,

(ii) $\partial\phi_2/\partial z=0$ on T,

(iii) $\phi_2=0$ on R,

(iv) $\partial\phi_2/\partial y$ is finite at the trailing edge of Σ (Kutta-Joukowski condition).

In order to solve this boundary-value problem, it is sufficient, as a first step, to determine ϕ_2 on Σ. The potential ϕ_2 is then known immediately on T from the condition that ϕ_2 is continuous on the trailing edge of Σ, and hence ϕ_2 is known everywhere on $y=+0$.

The potential ϕ_2 is then given everywhere in $y>0$ by (6.3.5) with $Q=0$, or by (6.3.6), and also in $y<0$, since ϕ_2 is an odd function of y. Alternatively, $\partial\phi_2/\partial z$ can be calculated on Σ, and ϕ_2 obtained from (6.3.7); this is probably the easier method.

The special case of (6.3.4) when $y_1=+0$ and $Q=0$ is

$$\phi_2(x_1,+0,z_1)=-\frac{1}{\pi}\iint\left(\frac{\partial\phi_2}{\partial y}\right)_{y=+0}\frac{dz\,dx}{\sqrt{\{(z-z_1)^2-B^2(x-x_1)^2\}}},\tag{6.5.1}$$

where the integration is over that part of $y=0$ for which

$$z\leqslant z_1-B\,|\,x-x_1\,|.\tag{6.5.2}$$

This integral can be transformed by putting

$$\left.\begin{aligned}\xi&=z_1-Bx_1, & \eta&=z_1+Bx_1,\\ \xi'&=z-Bx, & \eta'&=z+Bx,\end{aligned}\right\}\tag{6.5.3}$$

so that ξ,η are characteristic co-ordinates in the plane $y=0$, the lines $\xi=$ *constant* and $\eta=$ *constant* making the Mach angle with the stream direction. Then, since

$$\frac{\partial(z,x)}{\partial(\xi',\eta')}=\frac{1}{2B},\tag{6.5.4}$$

and by writing

$$N(\xi',\eta')=-\frac{1}{2\pi B}\left(\frac{\partial\phi_2}{\partial y}\right)_{y=+0},\quad \phi(\xi,\eta)=\phi(x_1,+0,z_1),\tag{6.5.5}$$

(6.5.1) becomes
$$\phi(\xi,\eta)=\iint\frac{N(\xi',\eta')\,d\xi'\,d\eta'}{\sqrt{\{(\xi-\xi')(\eta-\eta')\}}},\tag{6.5.6}$$

the integration being over the domain

$$\xi'\leqslant\xi,\quad \eta'\leqslant\eta.\tag{6.5.7}$$

The function $N(\xi',\eta')$ is not known on R and T in the influence domain of the wing, but can be determined there by setting up an integral equation which expresses the conditions (ii) and (iii) above. This procedure, combined with the transformation to characteristic co-ordinates, was used by Evvard (1947) to solve a restricted class of anti-symmetrical problems, and was subsequently extended independently by Evvard (1950), Hayes and Linstone (1948), and Ward (1949*c*) to cover more general classes of wings.

The above method has also been used by Goldsworthy (1952) to solve a very similar mixed problem for a symmetrical wing, in which the wing shape is to be found for a given pressure distribution. In this case $\partial\phi/\partial z$ is given on Σ, $\partial\phi/\partial y=0$ on the rest of the plane, and $\partial\phi/\partial y$ is to be determined on Σ; for details the reader is referred to the original paper.

Now let L be the boundary of the mean wing surface, and, as in fig. 6.3, let A, C, J, G be the points on L at which its tangents make the Mach angle with the stream direction, so that these tangents AA', etc., are characteristic lines in the plane of Σ. AG is then the supersonic leading edge of Σ. In fig. 6.3, BD, HK are lines parallel to the stream direction, tangent to L at B and H, which bound the linearized vortex sheet, T; AE, BB', CC', GE, HH', JJ' are all characteristic lines.

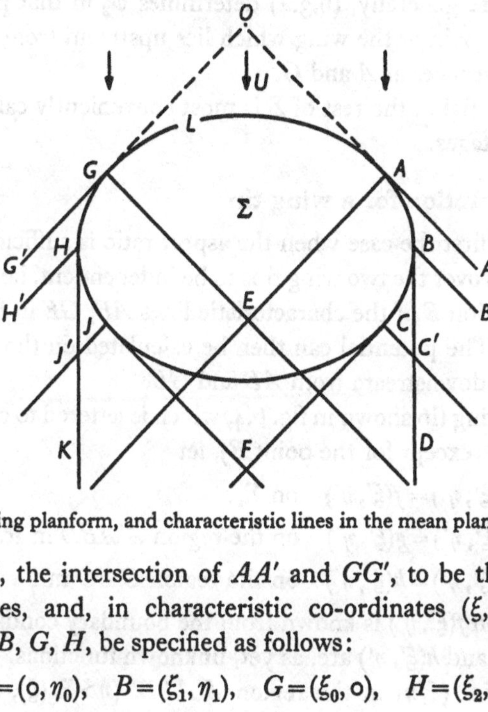

Fig. 6.3. Wing planform, and characteristic lines in the mean plane of the wing.

Take O, the intersection of AA' and GG', to be the origin of co-ordinates, and, in characteristic co-ordinates (ξ, η), let the points A, B, G, H, be specified as follows:

$$A = (0, \eta_0), \quad B = (\xi_1, \eta_1), \quad G = (\xi_0, 0), \quad H = (\xi_2, \eta_2).$$

Let the equations for L in its various parts be

$$GAC\colon\ \xi = H_1(\eta); \quad CJG\colon\ \xi = H_2(\eta);$$

or

$$JGA\colon\ \eta = \Xi_1(\xi); \quad ACJ\colon\ \eta = \Xi_2(\xi);$$

so that Ξ_1, Ξ_2, H_1, H_2 are single-valued functions on the arcs of L for which they are defined, if, for simplicity, the curvature of L is assumed to have the same sign everywhere, as shown in fig. 6.3.

If this condition is not satisfied, the following analysis may require modification, but this is easily carried out.

The potential in the region AGE of Σ may be calculated at once from (6.5.1) or (6.5.6). The potential and its derivatives vanish upstream from the envelope of the downstream Mach cones with vertices on the supersonic leading edge, AG, by condition (ix) of §4.5, and this envelope intersects $y=0$ in the curve $A'AGG'$. Hence $(\partial\phi_2/\partial y)_{y=0}=0$ upstream from $A'AGG'$, and it is known on the region AGE, so that, for a point (ξ,η) in AGE, $(\partial\phi_2/\partial y)_{y=+0}$ or $N(\xi',\eta')$ is known in the region of integration $(\xi'\leqslant\xi, \eta'\leqslant\eta)$ for (6.5.6). More generally, (6.3.4) determines ϕ_2 in that part of the influence domain of the wing which lies upstream from the Mach cones with vertices at A and G.

The potential in the rest of Σ is most conveniently calculated in successive stages.

6.6. The solution for a wing tip

Consider first the case when the aspect ratio is sufficiently large for the flows over the two wing tips to be independent, i.e. such that the intersection E of the characteristic lines AE, GE in fig. 6.3 lies outside Σ. The potential can then be calculated on those parts of Σ which lie downstream from AE and GE.

For the wing tip shown in fig. 6.4, which is lettered to correspond with fig. 6.3 (except for the point E), let

$$\left.\begin{aligned} N(\xi',\eta') &= f(\xi',\eta') \quad \text{on } \Sigma,\\ N(\xi',\eta') &= g(\xi',\eta') \quad \text{on the region } A'ABD \text{ in } R,\\ N(\xi',\eta') &= h(\xi',\eta') \quad \text{on the region } C'BC \text{ in } T. \end{aligned}\right\} \tag{6.6.1}$$

The function $f(\xi',\eta')$ is known from the boundary condition on Σ, but $g(\xi',\eta')$ and $h(\xi',\eta')$ are, as yet, unknown functions.

For a point (ξ,η) in the region $A'ABB'$ $(\eta>\Xi_2(\xi),\ 0<\xi<\xi_1)$, (6.5.6) and condition (iii) of §6.5 above give

$$\phi(\xi,\eta)=\int_0^\xi \frac{d\xi'}{\sqrt{(\xi-\xi')}}\left\{\int_{\Xi_2(\xi')}^{\eta}\frac{g(\xi',\eta')}{\sqrt{(\eta-\eta')}}\,d\eta'+\int_{\Xi_1(\xi')}^{\Xi_2(\xi')}\frac{f(\xi',\eta')}{\sqrt{(\eta-\eta')}}\,d\eta'\right\}=0. \tag{6.6.2}$$

This equation is satisfied if

$$\int_{\Xi_2(\xi)}^{\eta}\frac{g(\xi,\eta')}{\sqrt{(\eta-\eta')}}\,d\eta'+\int_{\Xi_1(\xi)}^{\Xi_2(\xi)}\frac{f(\xi,\eta')}{\sqrt{(\eta-\eta')}}\,d\eta'=0, \tag{6.6.3}$$

which is an integral equation of Abel's type for $g(\xi,\eta)$. The solution of (6.6.3) is

$$g(\xi,\eta) = -\frac{1}{\pi\sqrt{\{\eta-\Xi_2(\xi)\}}}\int_{\Xi_1(\xi)}^{\Xi_2(\xi)} \frac{f(\xi,\eta')\sqrt{\{\Xi_2(\xi)-\eta'\}}}{\eta-\eta'}\,d\eta', \quad (6.6.4)$$

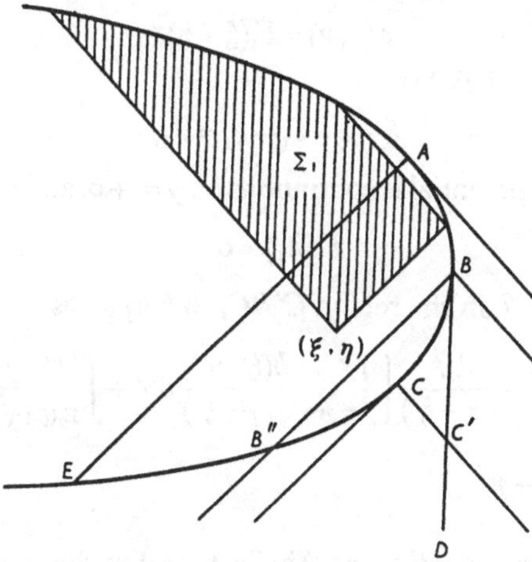

Fig. 6.4. Characteristic lines, and domain of integration in the expression (6.6.7) for the case of an independent wing-tip.

which gives $(\partial\phi_2/\partial y)_{y=+0}$ in the region $A'ABB'$ in terms of its value on Σ. The potential at a point (ξ,η) on Σ in the region $B''BAE$ $(H_1(\eta)<\xi<H_2(\eta),\ \eta_0<\eta<\eta_1)$ is

$$\phi(\xi,\eta) = \int_{H_1(\eta)}^{\xi} \frac{d\xi'}{\sqrt{(\xi-\xi')}}\int_{\Xi_1(\xi')}^{\eta} \frac{f(\xi',\eta')}{\sqrt{(\eta-\eta')}}\,d\eta'$$
$$+\int_0^{H_1(\eta)} \frac{d\xi'}{\sqrt{(\xi-\xi')}}\left\{\int_{\Xi_2(\xi')}^{\eta} \frac{g(\xi',\eta')}{\sqrt{(\eta-\eta')}}\,d\eta' + \int_{\Xi_1(\xi')}^{\Xi_2(\xi')} \frac{f(\xi',\eta')}{\sqrt{(\eta-\eta')}}\,d\eta'\right\}. \quad (6.6.5)$$

The last term in (6.6.5) vanishes by virtue of (6.6.3), and the final result for the value of the potential on Σ in the region $B''BAE$ is

$$\phi(\xi,\eta) = \int_{H_1(\eta)}^{\xi} \frac{d\xi'}{\sqrt{(\xi-\xi')}}\int_{\Xi_1(\xi')}^{\eta} \frac{f(\xi',\eta')}{\sqrt{(\eta-\eta')}}\,d\eta', \quad (6.6.6)$$

or, in terms of the usual cartesian co-ordinates, x, y, z,

$$\phi_2(x_1,+0,z_1) = -\frac{1}{\pi}\iint_{\Sigma_1}\left(\frac{\partial\phi_2}{\partial y}\right)_{y=+0} \frac{dz\,dx}{\sqrt{\{(z-z_1)^2-B^2(x-x_1)^2\}}}, \quad (6.6.7)$$

where Σ_1 is the area shown shaded in fig. 6.4. This is Evvard's result.

The potential for the region CBB'' must now be calculated in order to complete the solution. On T for $y=+0$, the potential is a function of x alone, say

$$\phi(\xi,\eta)=F_1(\xi-\eta). \tag{6.6.8}$$

If the equation of BD is

$$\xi-\eta=\xi_1-\eta_1=\lambda_1, \tag{6.6.9}$$

then, since the potential is continuous on $y=+0$, and vanishes on R,

$$F_1(\lambda_1)=0. \tag{6.6.10}$$

For a point on T in the region $C'BC$, (6.6.6) gives

$$\phi(\xi,\eta)=\int_{\eta+\lambda_1}^{\xi}\frac{d\xi'}{\sqrt{(\xi-\xi')}}\left\{\int_{\Xi_2(\xi')}^{\eta}\frac{h(\xi',\eta')}{\sqrt{(\eta-\eta')}}d\eta'+\int_{\Xi_1(\xi')}^{\Xi_2(\xi')}\frac{f(\xi',\eta')}{\sqrt{(\eta-\eta')}}d\eta'\right\}$$
$$=F_1(\xi-\eta). \tag{6.6.11}$$

This is another equation of Abel's type for the contents of the brackets, and its solution is, on taking account of (6.6.10),

$$\int_{\Xi_2(\xi)}^{\eta}\frac{h(\xi,\eta')}{\sqrt{(\eta-\eta')}}d\eta'+\int_{\Xi_1(\xi)}^{\Xi_2(\xi)}\frac{f(\xi,\eta')}{\sqrt{(\eta-\eta')}}d\eta'=\frac{1}{\pi}\int_{\eta+\lambda_1}^{\xi}\frac{F_1'(\xi'-\eta)}{\sqrt{(\xi-\xi')}}d\xi', \tag{6.6.12}$$

where F_1' is the first derivative of F_1. On solving for $h(\xi,\eta)$, (6.6.12) gives

$$h(\xi,\eta)=-\frac{1}{\pi\sqrt{\{\eta-\Xi_2(\xi)\}}}\int_{\Xi_1(\xi)}^{\Xi_2(\xi)}\frac{f(\xi,\eta')\sqrt{\{\Xi_2(\xi)-\eta'\}}}{\eta-\eta'}d\eta'$$
$$+\frac{1}{\pi^2}\frac{\partial}{\partial\eta}\int_{\Xi_2(\xi)}^{\eta}\frac{d\eta'}{\sqrt{(\eta-\eta')}}\int_{\eta'+\lambda_1}^{\xi}\frac{F_1'(\xi'-\eta')}{\sqrt{\{(\xi-\xi')}}d\xi'. \tag{6.6.13}$$

The function F_1 is determined from the Kutta-Joukowski condition that $h(\xi,\eta)$ must be finite at the subsonic trailing edge BC, where $\eta=\Xi_2(\xi)$. It will be seen that the second term in (6.6.13) must contain a term which cancels the singularity at the trailing edge contained in the first term. In fact, it is not necessary to solve for F_1 in order to determine the potential on the wing. Let a function $G_1(\zeta)$ be defined by

$$G_1(\zeta)=\frac{1}{\pi}\int_{\lambda_1}^{\zeta}\frac{F_1'(\zeta')}{\sqrt{(\zeta-\zeta')}}d\zeta'; \tag{6.6.14}$$

then the second term in (6.6.13) may be written

$$\frac{1}{\pi}\frac{\partial}{\partial\eta}\int_{\Xi_1(\xi)}^{\eta}\frac{G_1(\xi-\eta')}{\sqrt{(\eta-\eta')}}\,d\eta'=\frac{1}{\pi}\left\{\frac{G_1[\xi-\Xi_2(\xi)]}{\sqrt{\{\eta-\Xi_2(\xi)\}}}-\int_{\Xi_1(\xi)}^{\eta}\frac{G_1'(\xi-\eta')}{\sqrt{(\eta-\eta')}}\,d\eta'\right\}. \tag{6.6.15}$$

As the trailing edge is approached from a point in the region $C'BC$, $\eta\to\Xi_2(\xi)$, and the first term on the right-hand side of (6.6.15) becomes infinite, while the second term tends to zero, as a result of the continuity of the function F_1. Hence, by combining (6.6.13) and (6.6.15), it follows that, if $h(\xi,\eta)$ is bounded at $\eta=\Xi_2(\xi)$, then

$$G_1[\xi-\Xi_2(\xi)]=\int_{\Xi_1(\xi)}^{\Xi_2(\xi)}\frac{f(\xi,\eta')}{\sqrt{\{\Xi_2(\xi)-\eta'\}}}\,d\eta'. \tag{6.6.16}$$

Therefore, from (6.6.13), (6.6.15) and (6.6.16),

$$h(\xi,\eta)=\frac{1}{\pi}\sqrt{\{\eta-\Xi_2(\xi)\}}\int_{\Xi_1(\xi)}^{\Xi_2(\xi)}\frac{f(\xi,\eta')}{(\eta-\eta')\sqrt{\{\Xi_2(\xi)-\eta'\}}}\,d\eta' -\frac{1}{\pi}\int_{\Xi_1(\xi)}^{\eta}\frac{G_1'(\xi-\eta')}{\sqrt{(\eta-\eta')}}\,d\eta'. \tag{6.6.17}$$

The value of $h(\xi,\eta)$ at the trailing edge BC can be calculated. Let

$$J=\sqrt{\{\eta-\Xi_2(\xi)\}}\int_{\Xi_1(\xi)}^{\Xi_2(\xi)}\frac{\{f(\xi,\eta')-f[\xi,\Xi_2(\xi)]\}}{(\eta-\eta')\sqrt{\{\Xi_2(\xi)-\eta'\}}}\,d\eta', \tag{6.6.18}$$

and let $f(\xi,\eta)$ be such that $\lim_{\eta\to\Xi_2(\xi)} J=0.$ (6.6.19)

Then, since

$$\sqrt{\{\eta-\Xi_2(\xi)\}}\int_{\Xi_1(\xi)}^{\Xi_2(\xi)}\frac{d\eta'}{(\eta-\eta')\sqrt{\{\Xi_2(\xi)-\eta'\}}} =\pi-2\tan^{-1}\sqrt{\left[\frac{\eta-\Xi_2(\xi)}{\Xi_2(\xi)-\Xi_1(\xi)}\right]}, \tag{6.6.20}$$

and the second term of (6.6.17) tends to zero as $\eta\to\Xi_2(\xi)$, it follows from (6.6.17) that $\lim_{\eta\to\Xi_2(\xi)} h(\xi,\eta)=f[\xi,\Xi_2(\xi)].$ (6.6.21)

The condition (6.6.19) is certainly satisfied if $f(\xi,\eta)$ satisfies the conditions

(i) $|f(\xi,\eta)-f[\xi,\Xi_2(\xi)]|\leqslant K_1(\Xi_2(\xi)-\eta)^{\delta}$

for $\Xi_2(\xi)\geqslant\eta_1\geqslant k_1>\Xi_1(\xi)$,

(ii) $\int_{\Xi_1(\xi)}^{k_1}|f(\xi,\eta)-f[\xi,\Xi_2(\xi)]|\,d\eta\leqslant K_2$,

where K_1, K_2 and δ are positive constants; these conditions are more general than those usually required in any physical problem.

The result (6.6.21) shows that $\partial\phi_2/\partial y$ is continuous at the subsonic trailing edge, and therefore, from (6.5.6), $\partial\phi_2/\partial x$ and $\partial\phi_2/\partial z$ are also continuous there. Thus $\nabla\phi_2$ is continuous at the subsonic trailing edge.

The potential at a point (ξ, η) on Σ in the region CBB'' is, from (6.5.6),

$$\phi(\xi,\eta)=\int_{H_1(\eta)}^{\xi}\frac{d\xi'}{\sqrt{(\xi-\xi')}}\int_{\Xi_1(\xi')}^{\eta}\frac{f(\xi',\eta')}{\sqrt{(\eta-\eta')}}d\eta'$$
$$+\int_{\eta+\lambda_1}^{H_1(\eta)}\frac{d\xi'}{\sqrt{(\xi-\xi')}}\left\{\int_{\Xi_2(\xi')}^{\eta}\frac{h(\xi',\eta')}{\sqrt{(\eta-\eta')}}d\eta'+\int_{\Xi_1(\xi')}^{\Xi_2(\xi')}\frac{f(\xi',\eta')}{\sqrt{(\eta-\eta')}}d\eta'\right\}. \tag{6.6.22}$$

By using (6.6.12) to replace the contents of the brackets in the second term of (6.6.22), and remembering the definition of G_1 (6.6.14), the final expression for the potential on CBB'' is

$$\phi(\xi,\eta)=\int_{H_1(\eta)}^{\xi}\frac{d\xi'}{\sqrt{(\xi-\xi')}}\int_{\Xi_1(\xi')}^{\eta}\frac{f(\xi',\eta')}{\sqrt{(\eta-\eta')}}d\eta'+\int_{\eta+\lambda_1}^{H_1(\eta)}\frac{G_1(\xi'-\eta)}{\sqrt{(\xi-\xi')}}d\xi', \tag{6.6.23}$$

where G_1 is given in terms of $f(\xi, \eta)$ by (6.6.16). The first term in (6.6.23) is identical in form with that of (6.6.6) or (6.6.7); the second term represents the effect of that part of the vortex sheet, T, which lies in the dependence domain of the point (ξ, η).

On the subsonic trailing edge BC, the first term in (6.6.23) vanishes, and by substituting for G_1 from (6.6.14), and inverting the order of integration, it can be verified that the potential is continuous there. It can also be verified that $\partial\phi_2/\partial z = 0$ at BC, by differentiating (6.6.23), and using (6.6.16).

The potential is now known everywhere on this wing tip, and hence on the other wing tip, by using the formal symmetry of the problem with respect to ξ and η. The results are the same, but with reversed sign, on $y = -0$.

6.7. Independent subsonic edges

When the wing-tip flows are not independent in the sense of §6.6, an application of the above results gives the potential on the wing for the case when the characteristic lines AE, GE intersect on Σ, and intersect the supersonic trailing edge JC. In this case the flows over the subsonic edges are said to be independent, since neither subsonic edge lies in the influence domain of the other.

Consider the wing planform shown in fig. 6.5 (*a*), for which the characteristic lines BB'', HH'' intersect on Σ, at E'.

From the results of § 6.6, the potential is known on Σ except in the region $G''EA''$. Let the functions F_2, G_2 for the left-hand tip correspond respectively to the functions F_1, G_1 defined in § 6.6 for the right-hand tip, and let $\lambda_2 = \eta_2 - \xi_2$, so that the equation of the line HK is

$$\eta - \xi = \lambda_2. \tag{6.7.1}$$

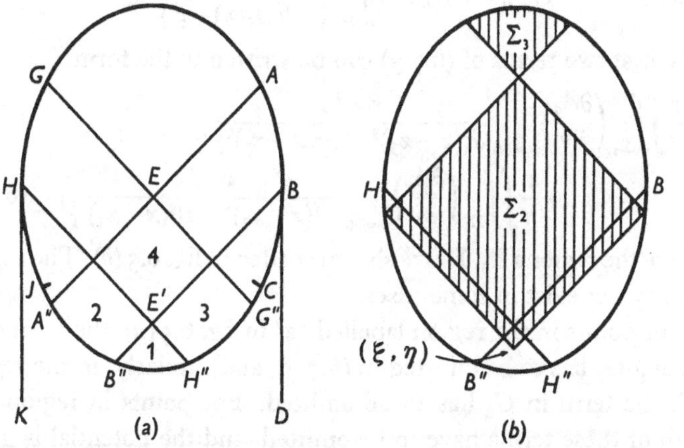

Fig. 6.5. (*a*) Wing planform and characteristic lines, and (*b*) domains of integration in the expression (6.7.5), for the case of independent subsonic edges.

Then $F_2(\lambda_2) = 0$, and the potential in the region, $H''E'B''$, say, can be written down immediately from (6.6.23) in the form

$$\begin{aligned}\phi(\xi, \eta) = &\int_{H_1(\eta)}^{\xi} \frac{\mathrm{d}\xi'}{\sqrt{(\xi - \xi')}} \left\{ \int_{\Xi_1(\xi)}^{\eta} \frac{f(\xi', \eta')}{\sqrt{(\eta - \eta')}} \mathrm{d}\eta' + \int_{\Xi_1(\xi')}^{\Xi_1(\xi)} \frac{f(\xi', \eta')}{\sqrt{(\eta - \eta')}} \mathrm{d}\eta' \right\} \\ &+ \int_{\eta + \lambda_1}^{H_1(\eta)} \frac{G_1'(\xi' - \eta)}{\sqrt{(\xi - \xi')}} \mathrm{d}\xi' + \int_{\xi_2}^{\xi} \frac{\mathrm{d}\xi'}{\sqrt{(\xi - \xi')}} \left\{ \int_{\xi' + \lambda_2}^{\Xi_1(\xi')} \frac{h(\xi', \eta')}{\sqrt{(\eta - \eta')}} \mathrm{d}\eta' \right. \\ &\left. + \int_0^{\xi' + \lambda_2} \frac{g(\xi', \eta')}{\sqrt{(\eta - \eta')}} \mathrm{d}\eta' \right\} + \int_0^{\xi_2} \frac{\mathrm{d}\xi'}{\sqrt{(\xi - \xi')}} \int_0^{\Xi_1(\xi')} \frac{g(\xi', \eta')}{\sqrt{(\eta - \eta')}} \mathrm{d}\eta'.\end{aligned} \tag{6.7.2}$$

By inverting the order of integration in the second repeated integral containing $f(\xi', \eta')$, and using results for the left-hand wing tip which correspond to (6.6.3) and (6.6.12) for the right-hand tip,

$g(\xi', \eta')$ and $h(\xi', \eta')$ can be eliminated from (6.7.2). This yields the result

$$\phi(\xi, \eta) = \int_{H_1(\eta)}^{\xi} \frac{d\xi'}{\sqrt{(\xi-\xi')}} \int_{\Xi_1(\xi)}^{\eta} \frac{f(\xi', \eta')}{\sqrt{(\eta-\eta')}} d\eta'$$

$$- \int_{H_1[\Xi_1(\xi)]}^{H_1(\eta)} \frac{d\xi'}{\sqrt{(\xi-\xi')}} \int_{\Xi_1(\xi')}^{\Xi_1(\xi)} \frac{f(\xi', \eta')}{\sqrt{(\eta-\eta')}} d\eta'$$

$$+ \int_{\eta+\lambda_1}^{H_1(\eta)} \frac{G_1(\xi'-\eta)}{\sqrt{(\xi-\xi')}} d\xi' + \int_{\xi+\lambda_2}^{\Xi_1(\xi)} \frac{G_2(\eta'-\xi)}{\sqrt{(\eta-\eta')}} d\eta', \quad (6.7.3)$$

where
$$G_2[\eta - H_2(\eta)] = \int_{H_1(\eta)}^{H_2(\eta)} \frac{f(\xi', \eta)}{\sqrt{\{H_2(\eta)-\xi'\}}} d\xi'. \quad (6.7.4)$$

The first two terms of (6.7.3) can be written in the form

$$-\frac{1}{\pi} \iint_{\Sigma_2} \left(\frac{\partial \phi_2}{\partial y}\right)_{y=+0} \frac{dz\, dx}{\sqrt{\{(z-z_1)^2 - B^2(x-x_1)^2\}}}$$

$$+\frac{1}{\pi} \iint_{\Sigma_3} \left(\frac{\partial \phi_2}{\partial y}\right)_{y=+0} \frac{dz\, dx}{\sqrt{\{(z-z_1)^2 - B^2(x-x_1)^2\}}}, \quad (6.7.5)$$

where the regions Σ_2, Σ_3 are shown shaded in fig. 6.5 (*b*). The region Σ_3 may not exist in some cases.

For points in the region labelled '2' in fig. 6.5 (*a*), the term containing G_1 has to be omitted in (6.7.3), and similarly, in the region '3', the term in G_2 has to be omitted. For points in region '4', both of these terms have to be omitted, and the potential is given by the expression (6.7.5); this result for such regions was given by Evvard (1947).

6.8. Lifting planes

The velocity components at points on the wing surface can be found from the expressions for the perturbation velocity potential, given in the preceding sections of this chapter, either by differentiating with respect to x or z, or by replacing ϕ by $\partial\phi/\partial x$ or $\partial\phi/\partial z$, since both these derivatives satisfy the equation for ϕ. In the latter cases, the boundary conditions on the wing surface are conditions on $\partial^2\phi/\partial x\,\partial y$ or $\partial^2\phi/\partial y\,\partial z$, and care must be taken to include terms arising from discontinuities in these quantities at the edges of the wing. The resulting formulae, which are rather complicated analytical expressions in the general case, have been given by Evvard (1950). However, in the particular case of lifting planes, for which the incidence, α, is constant over the wing, a remarkable simplification occurs, and the surface integrals reduce to line integrals along the supersonic leading edge.

When the wing-tip flows are independent in the sense of § 6.6, three cases arise as shown in fig. 6.6. For a point $(x_1, +0, z_1)$ on the wing outside the influence domain of the subsonic edges, if the forward characteristic lines through the point (x_1, z_1) in the mean

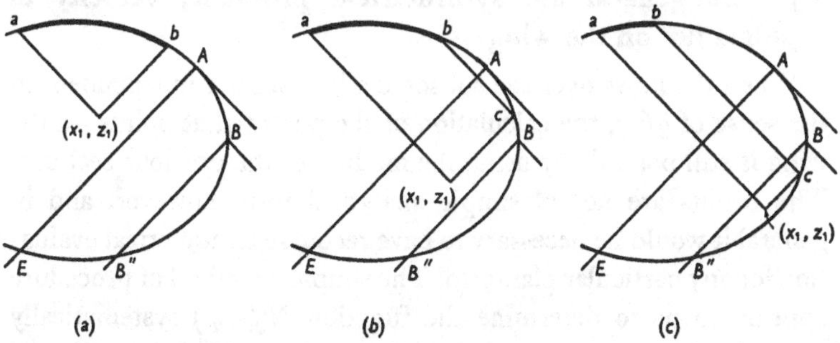

Fig. 6.6. Ranges of integration on the supersonic leading edge in the line integrals (6.8.1) and (6.8.2) for points in three different domains of the wing surface.

plane of the wing $(y=0)$ intersect the supersonic leading edge in the points a and b (see fig. 6.6 (a)), then

$$\left(\frac{\partial \phi_2}{\partial z}\right)_{y=+0} = \frac{U\alpha}{\pi}\int_{ab} \frac{dx}{\sqrt{\{(z-z_1)^2 - B^2(x-x_1)^2\}}}, \tag{6.8.1}$$

the integration being along the part ab of the supersonic leading edge. For a point in the region $ABB''E$ (fig. 6.6 (b)) if the forward characteristic lines from (x_1, z_1) intersect the supersonic leading edge in the point a and the subsonic leading edge in the point c, and if the forward characteristic line from the point c inside the wing planform intersects the supersonic leading edge in the point b (see fig. 6.6 (b)), then

$$\left(\frac{\partial \phi_2}{\partial z}\right)_{y_1=+0} = \frac{U\alpha}{\pi}\int_{ab} \frac{dx}{\sqrt{\{(z-z_1)^2 - B^2(x-x_1)^2\}}} + \frac{U\alpha}{\pi}\frac{1}{B+\tan\chi_c}\sqrt{\left\{\frac{z_1 - z_b + B(x_1 - x_b)}{z_1 - z_b - B(x_1 - x_b)}\right\}}, \tag{6.8.2}$$

where x_b, z_b are the co-ordinates of the point b, χ_c is the sweep of the subsonic leading edge at c, and the integration is along the edge ab as shown in fig. 6.6 (b). For a point in the region $BB''C$ (fig. 6.6 (c)), if the points a and b are defined as in the last case, then

$(\partial\phi_2/\partial z)_{y_1=+0}$ is given by (6.8.1) again when the Kutta-Joukowski condition is satisfied at the subsonic trailing edge. The expressions for $(\partial\phi_2/\partial x)_{y_1=+0}$ are similar.

Corresponding expressions can be found for the case of interacting wing-tip flows in the sense of §6.7.

6.9. The general anti-symmetrical problem; velocity at points not on the wing

When the flows over the subsonic edges are not independent in the sense of §6.7, the calculation of the potential at points on the wing is still possible by using the method of the previous sections. The results are not of simple analytical form however, and in general it would be necessary to have recourse to numerical evaluation for any particular planform. The simplest method of procedure appears to be to determine the function $N(\xi', \eta')$ systematically over the plane $y = +0$, and then to calculate $\phi(\xi, \eta)$ from (6.5.6). An indication of the steps required to carry out this procedure has been given by Ward (1949*c*), but no applications were made.

The method of this chapter fails to give useful results directly in the case when the wing has no supersonic leading edge, that is, when the wing is such that A and G (fig. 6.3) coincide at a point that may be called the vertex of the wing; for then Evvard's procedure leads to an infinite series for the potential. In this case, if the subsonic edges are straight, then the anti-symmetrical problem can be solved for special distributions of local incidence by the conical field method of the next chapter (see §7.5). By using solutions of this type, Evvard's method can be applied successfully to solve problems in which the subsonic edges are straight for a finite distance downstream from the vertex, since the process of solution, which would otherwise lead to an infinite series, can be terminated once the regions of integration off the wing consist only of points for which the conical field solution is known. Alternatively, if the curvature of the leading edges is finite in a neighbourhood of the vertex, then the conical field solution can be used to give an approximation to the sum of the later terms in the infinite series solution.

Once the potential is known all over the mean wing surface, it can be found everywhere else by using (6.3.5) or (6.3.7), the latter being preferable since the integration is over the wing only and it is not necessary to take 'finite parts'; the velocity components can then be found by differentiation. This method has been suggested by Ward (1949*b*), but Leslie (1952), in a paper summarizing the methods proposed for calculating downwash fields near the wakes of nearly plane wings, criticizes the method on the grounds that direct differentiation of (6.3.7) yields divergent integrals which have to be treated by Hadamard's 'finite-part' technique. As an alternative, Leslie suggests that (6.3.7) be integrated by parts before differentiation, which gives the velocity components in terms of line integrals along the leading edge and surface integrals over Σ; however, in general these integrals also diverge due to infinite values of $\partial\phi/\partial z$ and $\partial^2\phi/\partial x\,\partial z$, etc., at subsonic leading edges, and again a 'finite-part' method has to be used or else further transformations have to be made, but when the wind has no subsonic leading edges, as for a rectangular planform, the difficulty disappears. An alternative approach to the problem has been given by Robinson and Hunter-Tod (1947*a*), who replace the lifting wing by an equivalent system of bound vortices and a trailing vortex sheet, whose strengths are given in terms of the potential at the wing surface by means of equation (3.8.4). The resulting formulae for the velocity components are rather complicated expressions for which the reader is referred to the original papers.

The velocity potential in those parts of the dependence domains of the subsonic leading edges that do not include any part of the dependence domains of the subsonic trailing edges has been obtained directly by Friedlander (1951),† who applies the identity (3.6.1) to suitably chosen domains. The expression thus obtained for the potential in $y_1 > 0$, when the wing-tip flows are independent, is

$$\phi_2(\mathbf{R}_1) = -\frac{1}{\pi}\iint_{\Sigma_1}\left(\frac{1}{R_B}\frac{\partial\phi_2}{\partial y}\right)_{y=+0} dz\,dx, \qquad (6.9.1)$$

where Σ_1 is the area of wing planform bounded by the upstream branch of the hyperbola

$$B^2(x-x_1)^2 + B^2y_1^2 = (z-z_1)^2, \qquad (6.9.2)$$

the upstream characteristic line inside the wing planform from the intersection of the hyperbola (6.9.2) and the subsonic leading

† Friedlander considers the problem in terms of the diffraction of sound, but the analysis is identical with that for the wing problem in this case.

edge, and the supersonic leading edge. This is a natural generalization of Evvard's result (6.6.7). A similar formula is easily written down for the case of independent subsonic edges, being the same generalization of (6.7.5). A corresponding formula for points in the dependence domains of the subsonic trailing edges, and including the effects of the trailing vortex sheet, has not been developed at the time of writing.

6.10. Aerodynamic forces

The aerodynamic forces can be calculated from the results of §§4.6 and 4.7. If ϕ_1 and ϕ_2 are respectively the symmetrical and anti-symmetrical potentials, then, on taking account of their dependence on y, the contribution to the drag force from Σ in (4.6.17) can easily be shown to be

$$-2\rho_0 \iint_\Sigma \left(\frac{\partial\phi_1}{\partial y}\frac{\partial\phi_1}{\partial z}+\frac{\partial\phi_2}{\partial y}\frac{\partial\phi_2}{\partial z}\right)_{y=+0} dz\,dx, \qquad (6.10.1)$$

where the integration is over one side ($y=+0$) of Σ only.

The contribution to the drag from the edge forces can be determined from the results of §4.7. When the subsonic leading edges are independent, the function K in (4.7.3) and (4.7.5) can be found by using the expression for $g(\xi,\eta)$, given by (6.6.4) near to AB, and the similar expression which is valid near GH in fig. 6.3, since there is no contribution from ϕ_1. Near to the edge AB, as $\eta\to\Xi_2(\xi)$,

$$\left(\frac{\partial\phi_2}{\partial y}\right)_{y=0} = -2\pi Bg(\xi,\eta) \sim \frac{2B}{\sqrt{\{\eta-\Xi_2(\xi)\}}}\int_{\Xi_1(\xi)}^{\Xi_2(\xi)} \frac{f(\xi,\eta')}{\sqrt{\{\Xi_2(\xi)-\eta'\}}}\,d\eta'. \qquad (6.10.2)$$

If x' is normal distance outwards from the edge of Σ and in the plane $y=0$, at a point where the sweep is χ ($\frac{1}{2}\pi-\mu<\chi<\frac{1}{2}\pi$ for a subsonic leading edge), then

$$\eta-\Xi_2(\xi) = -2x'\cos\mu\sec(\mu+\chi), \qquad (6.10.3)$$

and (4.7.3) gives

$$K = B\sqrt{\left(-\frac{2\cos(\mu+\chi)}{\cos\mu}\right)}\int_{\Xi_1(\xi)}^{\Xi_2(\xi)} \frac{f(\xi,\eta')}{\sqrt{\{\Xi_2(\xi)-\eta'\}}}\,d\eta'; \quad (6.10.4)$$

the suction force per unit length of edge is, (4.7.5),

$$\frac{\pi\rho_0 K^2}{\sqrt{(1-M^2\cos^2\chi)}}. \qquad (6.10.5)$$

A similar result can be obtained for the other subsonic leading edge. If the edges are rounded, then terms like (4.7.6) must be included to give a total outward force, F_e say, per unit length of edge.

Thus the total drag force on the wing is

$$D = -2\rho_0 \iint_\Sigma \left(\frac{\partial\phi_1}{\partial y}\frac{\partial\phi_1}{\partial z} + \frac{\partial\phi_2}{\partial y}\frac{\partial\phi_2}{\partial z}\right)_{y=+0} \mathrm{d}z\,\mathrm{d}x - \int_{JG,AC} F_e\,\mathrm{d}x, \tag{6.10.6}$$

the integrations being over the upper side of Σ ($y = +0$) and the subsonic edges respectively.

The lift force (parallel to the y-axis) is calculated most easily from (4.6.15), and is

$$2\rho_0 U \int_{HJCB} (\phi_2)_{y=+0}\,\mathrm{d}x, \tag{6.10.7}$$

where the integration is along the trailing edge of Σ, or along any line spanning the linearized vortex sheet, T.

The pitching, rolling and yawing moments can be calculated in a similar way, by using the results of §4.8.

6.11. The lift force on a wing with a straight supersonic trailing edge and no subsonic edges

The result (6.10.7) can be used to obtain a useful formula for the lift force on wings which have no subsonic edges, and whose trailing edge is straight and swept at an angle $\chi < \frac{1}{2}\pi - \mu$.†

Since the edges of the mean surface Σ are all supersonic edges, the flows near the upper and lower sides of Σ are independent and can be calculated separately. If the direction cosines of the outward normals to the upper and lower sides of Σ^* are respectively l_+, m_+, n_+ and l_-, m_-, n_-, then the boundary conditions on Σ for ϕ_2 are

$$\left(\frac{\partial\phi_2}{\partial y}\right)_{y=+0} = \left(\frac{\partial\phi_2}{\partial y}\right)_{y=-0} = -\tfrac{1}{2}U(n_+ - n_-). \tag{6.11.1}$$

The potential $\phi(x_1, +0, z_1)$ at a point (x_1, z_1) on the upper side of the trailing edge is given by (6.5.1), and $(\partial\phi_2/\partial y)_{y=+0}$ is known over the whole domain of integration, being given by (6.11.1) downstream from the leading edge, and being zero upstream from the leading edge. Hence, from (6.10.7), the lift force is

$$\frac{\rho_0 U^2}{\pi} \int_{\text{t.e.}} \mathrm{d}x_1 \iint_\Sigma \frac{(n_+ - n_-)\,\mathrm{d}z\,\mathrm{d}x}{\sqrt{\{(z-z_1)^2 - B^2(x-x_1)^2\}}}, \tag{6.11.2}$$

† Lagerstrom and Van Dyke (1949), Ward (1949c).

where the integrations are over Σ and the trailing edge of the wing. The order of integration can be inverted to give

$$\frac{\rho_0 U^2}{\pi} \iint_\Sigma (n_+ - n_-)\, dz\, dx \int_{\text{t.e.}} \frac{dx_1}{\sqrt{\{(z-z_1)^2 - B^2(x-x_1)^2\}}}; \tag{6.11.3}$$

the inner integral can be evaluated at once, and is

$$\frac{\pi \cos \chi}{\sqrt{(M^2 \cos^2 \chi - 1)}}, \tag{6.11.4}$$

a quantity independent of x and z. It follows that the lift coefficient, with respect to $\frac{1}{2}\rho_0 U^2$ and wing area, is

$$C_L = \frac{4 \cos \chi}{\sqrt{(M^2 \cos^2 \chi - 1)}} \overline{\tfrac{1}{2}(n_+ - n_-)}, \tag{6.11.5}$$

where the bar denotes the average value with respect to area.

This formula can also be deduced directly from the flow-reversal result (4.11.4) when Σ is defined as above, by taking

$$\mathbf{k}.\mathbf{n} = \tfrac{1}{2}(n_+ - n_-) \quad \text{and} \quad \mathbf{k} \wedge \boldsymbol{\alpha}_1 . \boldsymbol{\nu} = \textit{constant}.$$

Then, on using the appropriate expression for p_1, and taking Σ_1 to be the whole of Σ, (6.11.5) follows. When Σ_1 is not the whole of Σ, a generalization of (6.11.5) is obtained. By removing the restriction that Σ is a plane, and also by using (4.11.9) in a similar way, it is possible to obtain many other results of a similar character.

CHAPTER 7

CONICAL FIELDS IN SUPERSONIC FLOW

7.1. The general solution for the velocity components

Many problems in supersonic flow are such that no fundamental length can be specified (e.g. for flow past a semi-infinite cone or a semi-infinite plane wing with straight leading edges); the velocity is then constant on straight lines through a fixed point, called the vertex of the flow field. If this point is taken as the origin of cartesian co-ordinates x, y, z, then the components, u, v, w, of the perturbation velocity are homogeneous functions of degree zero in x, y, z. Such velocity fields are called conical fields, and their existence was first recognized by Busemann (1935), who subsequently developed the linearized theory of these flows (1943). As a consequence of the independence of the flows in suitably chosen regions of a complete supersonic flow field, it is possible for some parts of a given field to be of the conical type. These parts of flow fields can be treated independently by conical field theory, and the results superposed to give the flow in more extended regions (see §7.6).

The linearized theory of supersonic conical fields has been developed in detail by a number of writers, nearly all by different methods. The following treatment follows that of Goldstein and Ward (1950); a very similar treatment was given by Hayes (1946*a*). For an example of a different method, the paper by Lagerstrom (1948) may be consulted; this paper develops Busemann's original treatment (1943) in considerable detail.

For irrotational flow with no sources, the perturbation velocity $\mathbf{v}$ satisfies the equation $\nabla \wedge (\nabla \wedge \mathbf{v}) = 0$; when expressed in terms of rectangular cartesian co-ordinates, this equation becomes

$$\frac{\partial^2 \mathbf{v}}{\partial x^2} + \frac{\partial^2 \mathbf{v}}{\partial y^2} - B^2 \frac{\partial^2 \mathbf{v}}{\partial z^2} = 0. \tag{7.1.1}$$

On writing $$\frac{Bx}{z} = x_1 = r_1 \cos\theta, \quad \frac{By}{z} = y_1 = r_1 \sin\theta, \tag{7.1.2}$$

it follows from the fact that $\mathbf{v}$ is a homogeneous function of degree zero in x, y, z, that $\mathbf{v}$ is a function of x_1, y_1, or r_1, θ only, and with

r_1, θ as independent variables, (7.1.1) becomes

$$r_1^2(r_1^2-1)\frac{\partial^2\mathbf{v}}{\partial r_1^2}+r_1(2r_1^2-1)\frac{\partial\mathbf{v}}{\partial r_1}-\frac{\partial^2\mathbf{v}}{\partial\theta^2}=0. \qquad (7.1.3)$$

This equation is hyperbolic for $r_1>1$ and elliptic for $r_1<1$, and it can be simplified considerably by making further transformations of the independent variables.

For $r_1>1$, on putting

$$1/r_1=\cos\sigma, \qquad (7.1.4)$$

then (7.1.3) becomes

$$\frac{\partial^2\mathbf{v}}{\partial\sigma^2}-\frac{\partial^2\mathbf{v}}{\partial\theta^2}=0, \qquad (7.1.5)$$

which has the general solution

$$\mathbf{v}=\mathbf{f}(\theta+\sigma)+\mathbf{F}(\theta-\sigma). \qquad (7.1.6)$$

As r_1 increases from 1 to ∞, σ increases from 0 to $\frac{1}{2}\pi$. If $\mathbf{v}$ is to be single-valued, then $\mathbf{f}$ and $\mathbf{F}$ must be periodic functions of period 2π. $\mathbf{f}$ and $\mathbf{F}$ must be independent functions, since each depends on only one of the characteristic variables $\theta+\sigma$, $\theta-\sigma$, so both must be irrotational vectors if $\mathbf{v}$ is to be irrotational. For $\mathbf{f}$, this condition is

$$\nabla\wedge\mathbf{f}(\theta+\sigma)=-\mathbf{f}'(\theta+\sigma)\wedge\nabla(\theta+\sigma)=0, \qquad (7.1.7)$$

so $\mathbf{f}'(\theta+\sigma)$ must be parallel to $\nabla(\theta+\sigma)$, and a little algebra shows that $\nabla(\theta+\sigma)$ is parallel to a vector $\mathbf{\Lambda}(\theta+\sigma)$, where, in rectangular components,

$$\mathbf{\Lambda}(\lambda)=[\cos\lambda,\ \sin\lambda,\ -1/B]. \qquad (7.1.8)$$

Similarly it can be shown that $\mathbf{F}'(\theta-\sigma)$ must be parallel to $\mathbf{\Lambda}(\theta-\sigma)$. Hence the general solution for $\mathbf{v}$ when $r_1>1$ is

$$\mathbf{v}=\int^{\theta+\sigma}f(\lambda)\,\mathbf{\Lambda}(\lambda)\,d\lambda+\int^{\theta-\sigma}F(\lambda)\,\mathbf{\Lambda}(\lambda)\,d\lambda, \qquad (7.1.9)$$

where $f(\lambda)$ and $F(\lambda)$ are periodic functions of period 2π satisfying

$$\int_0^{2\pi}\{f(\lambda)+F(\lambda)\}\,\mathbf{\Lambda}(\lambda)\,d\lambda=0, \qquad (7.1.10)$$

since $\mathbf{v}$ is single-valued. In later applications it will be found convenient to work in terms of the rectangular components, f_1, f_2, f_3, of $\mathbf{f}$ and F_1, F_2, F_3, of $\mathbf{F}$. f_1, f_2, f_3 are functions of $\theta+\sigma$ only, and F_1, F_2, F_3 are functions of $\theta-\sigma$ only.

For $r_1<1$, on putting

$$s=-\cosh^{-1}(1/r_1)=\log\frac{1-\sqrt{(1-r_1^2)}}{r_1}, \qquad (7.1.11)$$

then (7.1.3) becomes $$\frac{\partial^2 \mathbf{v}}{\partial s^2}+\frac{\partial^2 \mathbf{v}}{\partial \theta^2}=0. \tag{7.1.12}$$

Now let $$\zeta=e^{s+i\theta}=\frac{1-\sqrt{(1-r_1^2)}}{r_1}e^{i\theta}=\frac{r_1}{1+\sqrt{(1-r_1^2)}}e^{i\theta}. \tag{7.1.13}$$

Then when $r_1=1$, $s=0$ and $e^s=1$; as r_1 decreases from 1 to 0, $1/r_1$ increases from 1 to ∞, s decreases from 0 to $-\infty$, and e^s decreases from 1 to 0. Hence the interior of the unit circle $r_1=1$ in the (r_1, θ)-plane corresponds with the interior of the unit circle $|\zeta|=1$ in the ζ-plane; the boundaries correspond, and θ is the same in both planes. In terms of the complex variable ζ, the general solution for $\mathbf{v}$ satisfying (7.1.12) is

$$\mathbf{v}=\Re\mathbf{g}(\zeta)=\Re[g_1(\zeta), g_2(\zeta), g_3(\zeta)], \tag{7.1.14}$$

where g_1, g_2, g_3 are analytic functions of ζ. The condition that $\mathbf{v}$ should be irrotational is that $\mathbf{g}'(\zeta)$ must be parallel to $\nabla\zeta$, and again after a little algebra, $\nabla\zeta$ can be shown to be parallel to a (complex) vector $\mathbf{K}(\zeta)$, where in rectangular components

$$\mathbf{K}(\zeta)=[-(\zeta^2+1), i(\zeta^2-1), 2\zeta/B]. \tag{7.1.15}$$

Hence the general solution for $\mathbf{v}$ when $r_1<1$ is

$$\mathbf{v}=\Re\int^{\zeta} g(\kappa)\mathbf{K}(\kappa)\,d\kappa, \tag{7.1.16}$$

where $g(\kappa)$ is an analytic function of κ. The rectangular components of $\mathbf{g}$ satisfy

$$g_1'(\zeta):g_2'(\zeta):g_3'(\zeta)=-(\zeta^2+1):i(\zeta^2-1):2\zeta/B, \tag{7.1.17}$$

which, together with (7.1.14), is equivalent to (7.1.16) and is found to be the most convenient form for the solution in applications.

If the interior of the unit circle $|\zeta|=1$ is transformed conformally into some region of a complex t-plane, say, by the transformation $\zeta=h(t)$, then the general solution of (7.1.12) can be written

$$\mathbf{v}=\Re\mathbf{G}(t)=\Re[G_1(t), G_2(t), G_3(t)], \tag{7.1.18}$$

and since $$\nabla t=(dt/d\zeta)\nabla\zeta, \tag{7.1.19}$$

it follows that $\mathbf{G}'(t)$ must be parallel to $\mathbf{K}(\zeta)$, or

$$G_1'(t):G_2'(t):G_3'(t)=-(h^2+1):i(h^2-1):2h/B. \tag{7.1.20}$$

This generalization of (7.1.17) facilitates the use of conformal transformation methods in the solution of boundary-value problems.

The above solution for $\mathbf{v}$ in a conical field can be derived in an alternative way that is not without interest. First it must be pointed out that the vector $\mathbf{\Lambda}(\lambda)$ in (7.1.8) satisfies the equation

$$\mathbf{\Lambda}.\mathbf{\Psi}.\mathbf{\Lambda}=0, \qquad (7.1.21)$$

so $\mathbf{\Lambda}$ satisfies the same equation as the normal to a characteristic surface, and hence the plane

$$\mathbf{R}.\mathbf{\Lambda}=x\cos\lambda+y\sin\lambda-z/B=\text{constant} \qquad (7.1.22)$$

is a characteristic plane. It follows that any vector function of $\mathbf{R}.\mathbf{\Lambda}$ satisfies (7.1.1), and if $\mathbf{v}$ is to be an irrotational vector then it must have the form

$$\mathbf{v}=f[\mathbf{R}.\mathbf{\Lambda}(\lambda)]\,\mathbf{\Lambda}(\lambda). \qquad (7.1.23)$$

This is true for any (real or complex) value of λ, hence

$$\mathbf{v}=\int f(\mathbf{R}.\mathbf{\Lambda},\lambda)\,\mathbf{\Lambda}(\lambda)\,d\lambda \qquad (7.1.24)$$

also gives a possible irrotational flow when the limits of integration are independent of the space variables, and sometimes when they are not.

A velocity field of this type which is a homogeneous function of degree n in the space variables is

$$\mathbf{v}=\int_C (\mathbf{R}.\mathbf{\Lambda})^n f_n(\lambda)\,\mathbf{\Lambda}(\lambda)\,d\lambda, \qquad (7.1.25)$$

which satisfies (7.1.1) if $n\geqslant 0$, and if C is a contour in a complex λ-plane which joins the points λ_1, λ_2, where λ_1, λ_2 satisfy

$$\mathbf{R}.\mathbf{\Lambda}(\lambda_{1,2})=0, \qquad (7.1.26)$$

or which joins either λ_1 or λ_2 to some fixed point in the plane. In terms of r_1, θ as defined in (7.1.2), (7.1.26) gives

$$\left.\begin{aligned}\lambda_{1,2}&=\theta\mp\cos^{-1}(1/r_1) \quad \text{for} \quad r_1\geqslant 1,\\ \lambda_{1,2}&=\theta\mp i\cosh^{-1}(1/r_1) \quad \text{for} \quad r_1\leqslant 1,\end{aligned}\right\} \qquad (7.1.27)$$

or, in terms of σ and s,

$$\lambda_{1,2}=\theta\mp\sigma \text{ for } r_1\geqslant 1, \quad \lambda_{1,2}=\theta\pm is \text{ for } r_1\leqslant 1. \qquad (7.1.28)$$

With these values of λ_1, λ_2, (7.1.25) can be written

$$\mathbf{v}=\int_{\lambda_1}^{\lambda_2} (\mathbf{R}.\mathbf{\Lambda})^n f_n(\lambda)\,\mathbf{\Lambda}(\lambda)\,d\lambda, \qquad (7.1.29)$$

where, if $\mathbf{v}$ is to be real, $f_n(\lambda)$ must satisfy the reality condition

$$f_n(\lambda)=-\bar{f}_n(\lambda), \qquad (7.1.30)$$

the bar denoting the complex conjugate; this shows that $f_n(\lambda)$ is purely imaginary when λ is real. $f_n(\lambda)$ is of course a periodic function of period 2π if $\mathbf{v}$ is single-valued.

For the special case $n=0$, (7.1.29) gives a conical field of the type described above, in which

$$\mathbf{v}=\int_{\lambda_1}^{\lambda_2} f_0(\lambda)\,\mathbf{\Lambda}(\lambda)\,d\lambda. \tag{7.1.31}$$

The connexion between (7.1.31) and (7.1.9) for $r_1 \geqslant 1$ is clear immediately. The equivalence of (7.1.31) and (7.1.16) for $r_1 \leqslant 1$ is made apparent by using the transformation

$$\kappa=e^{i\lambda}, \quad \zeta=e^{i\lambda_2}, \quad f_0(\lambda)=-i\,e^{2i\lambda}g(e^{i\lambda}), \tag{7.1.32}$$

and the reality condition (7.1.30). This provides a connexion between the functions $f(\lambda)$ and $F(\lambda)$ in (7.1.9) and the function $g(\kappa)$ in (7.1.16), and shows that the two former are not entirely independent.

This alternative derivation is not as instructive physically as that given first, particularly in the case of the flow in $r_1 \geqslant 1$, but it has the advantage that it shows that the velocity is continuous at the Mach cone of the vertex (corresponding to $r_1=1$), a result that is not easy to establish by direct use of (7.1.9) and (7.1.16). It also leads to an obvious generalization to 'homogeneous fields' for which the velocity components are all of the form *constant* $\times R^n$ on lines passing through the vertex of the field, where R is radial distance from the vertex. Conical fields are the special case of homogeneous fields for which $n=0$. The perturbation velocity for a homogeneous field of degree n is given by (7.1.29).

The theory of homogeneous fields has been developed by Germain (1949*b*), and Lomax and Heaslet (1951), starting from a general solution of different form from (7.1.29), and by Hayes, Roberts and Haaser (1952) directly from (7.1.29). For details of this work the reader is referred to the original papers.

7.2. Flow outside the Mach cone of the vertex

On the assumptions of linearized theory, the only allowable bodies in conical flows are those whose mean surfaces are planes which pass through the axis of the conical field, and which have straight edges which pass through the vertex of the field (that is, the origin of co-ordinates). As an example of such a body, consider a plane triangular wing at incidence α, whose mean surface is part of the plane $y=0$, which corresponds to a certain portion of the line $y_1=0$ in the (x_1, y_1)-plane. This portion may or may not lie, wholly or partially, outside the circle $r_1=1$, which corresponds to the Mach cone of the vertex; the case when it does is considered in this section. The boundary condition is $v=-U\alpha$ when $y=0$ on the wing, or when $y_1=0$ on the section of the wing in the (x_1, y_1)-plane.

The part of the (x_1, y_1)-plane for which $r_1 \geqslant 1$ corresponds in the (σ, θ)-plane to the strip $0 \leqslant \sigma \leqslant \frac{1}{2}\pi$. The axis $y_1=0$ corresponds to $\theta=0$ (for $x_1>0$) and $\theta=\pi$ (for $x_1<0$), so the boundaries in the (σ, θ)-plane, corresponding to those parts of the plane triangular wing that are outside the Mach cone, lie along $\theta=0$ and $\theta=\pi$ (and are repeated at intervals of 2π). They extend from $\sigma=0$ $(r_1=1)$, if the wing passes through the Mach cone, but do not reach to $\sigma=\frac{1}{2}\pi$ if the wing lies entirely downstream from the plane $z=0$, as is assumed here.

If the wing extends from $\sigma=0$ to $\sigma=\sigma_0$ on $\theta=0$, and to $\sigma=\sigma_1$ on $\theta=\pi$, then the case is that of a plane triangular wing whose edges are outside and on opposite sides of the Mach cone, and are swept back at angles χ_0, χ_1 $(\chi_0, \chi_1<\frac{1}{2}\pi-\mu)$, as in fig. 7.1 (*a*), where

$$\sec\sigma_0 = B\cot\chi_0, \quad \sec\sigma_1 = B\cot\chi_1. \tag{7.2.1}$$

Now near $\sigma=\frac{1}{2}\pi$,

$$\mathbf{v} = \frac{\partial \mathbf{v}}{\partial \sigma} = \frac{\partial \mathbf{v}}{\partial \theta} = 0, \tag{7.2.2}$$

since there is no disturbance beyond some finite value of r_1. Hence from (7.1.6),

$$\begin{aligned} \mathbf{f}(\theta+\sigma)+\mathbf{F}(\theta-\sigma) &= \mathbf{f}'(\theta+\sigma)-\mathbf{F}'(\theta-\sigma) \\ &= \mathbf{f}'(\theta+\sigma)+\mathbf{F}'(\theta-\sigma)=0, \end{aligned} \tag{7.2.3}$$

from which

$$\mathbf{f}'(\theta+\sigma) = \mathbf{F}'(\theta-\sigma) = 0, \tag{7.2.4}$$

and

$$\mathbf{f}(\theta+\sigma) = -\mathbf{F}(\theta-\sigma) = \mathbf{C} \tag{7.2.5}$$

near $\sigma=\frac{1}{2}\pi$, where $\mathbf{C}$ is a constant. Now $\mathbf{f}(\theta+\sigma)$ and $\mathbf{F}(\theta-\sigma)$ are constant along the lines $\theta+\sigma=$ *constant* and $\theta-\sigma=$ *constant* respectively. (These lines are characteristics of the differential equation (7.1.5).) If lines $\theta+\sigma=$ *constant* and $\theta-\sigma=$ *constant* can be drawn from any point, P, to $\sigma=\frac{1}{2}\pi$ without intersecting the cuts from $\sigma=0$ to $\sigma=\sigma_0$ along $\theta=0$ and from $\sigma=0$ to $\sigma=\sigma_1$ along

$\theta = \pi$ which represent the wing, then (7.2.5) holds, and $\mathbf{v} = 0$ at P. Thus $\mathbf{v}$ vanishes everywhere for $r_1 \geqslant 1$, except in the regions labelled 1, 2, 3, 4 in fig. 7.2, which correspond to the similarly labelled regions in fig. 7.1 (*b*), as is shown below.

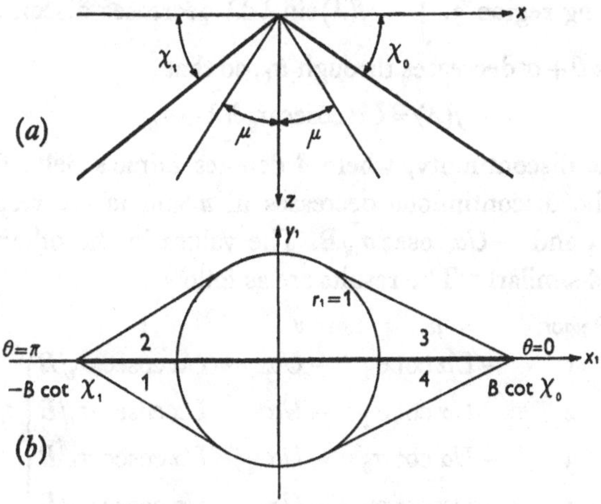

Fig. 7.1. (*a*) Wing leading edges in the (z, x)-plane, and (*b*) trace of the wing in the (x_1, y_1)-plane, when both leading edges are outside and on opposite sides of the Mach cone of the vertex.

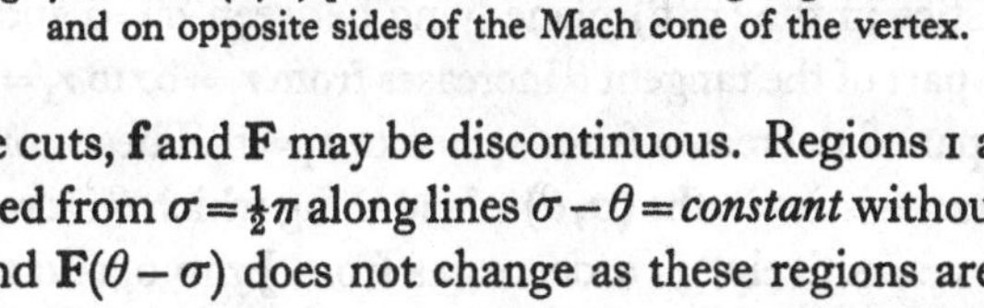

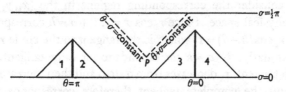

Fig. 7.2. The (σ, θ)-plane when both leading edges are outside and on opposite sides of the Mach cone of the vertex.

At the cuts, **f** and **F** may be discontinuous. Regions 1 and 3 may be reached from $\sigma = \frac{1}{2}\pi$ along lines $\sigma - \theta = constant$ without crossing a cut, and $\mathbf{F}(\theta - \sigma)$ does not change as these regions are entered; but the regions cannot be entered from $\sigma = \frac{1}{2}\pi$ along lines $\theta + \sigma = constant$ if a cut is not crossed, so $\mathbf{f}(\theta + \sigma)$ may change discontinuously. In order that v should be equal to $-U\alpha$ on the cuts, $f_2(\theta + \sigma)$ must decrease discontinuously by $U\alpha$ as $\theta + \sigma$ decreases through σ_0 when region 3 is entered, and as $\theta + \sigma$ decreases through $\pi + \sigma_1$ when region 1 is entered. Similarly, as regions 2 and 4 are entered, $f(\theta+\sigma)$ does not change, but $F_2(\theta-\sigma)$ decreases discontinuously by $U\alpha$ as $\theta-\sigma$ increases through $-\sigma_0$ and $\pi-\sigma_1$ respectively. Hence $v=-U\alpha$ throughout all four regions. The other two components, u and w, can be found from (7.1.9). For example, on entering region 3, $\int^{\sigma+\theta} f(\lambda)\sin\lambda\, d\lambda$ decreases discontinuously by $U\alpha$ as $\theta+\sigma$ decreases through σ_0, so that

$$f(\lambda)=U\alpha \operatorname{cosec}\sigma_0\,\delta(\lambda-\sigma_0)$$

from this discontinuity, where δ denotes Dirac's delta function. Hence the discontinuous decreases in u and w are respectively $U\alpha\cot\sigma_0$ and $-U\alpha\operatorname{cosec}\sigma_0/B$. The values in the other regions are found similarly. The results are as follows:

$$\left.\begin{array}{cccc}
\text{Region} & u & v & w \\
1 & -U\alpha\cot\sigma_1 & -U\alpha & -U\alpha\operatorname{cosec}\sigma_1/B \\
2 & U\alpha\cot\sigma_1 & -U\alpha & U\alpha\operatorname{cosec}\sigma_1/B \\
3 & -U\alpha\cot\sigma_0 & -U\alpha & U\alpha\operatorname{cosec}\sigma_0/B \\
4 & U\alpha\cot\sigma_0 & -U\alpha & -U\alpha\operatorname{cosec}\sigma_0/B
\end{array}\right\} \qquad (7.2.6)$$

In these results v is an even function, and u and w are odd functions of y, as they should be.

Now consider the corresponding regions in the (x_1, y_1)-plane and in physical space. Since $r_1\cos\sigma=1$, $\theta+\sigma=k$ corresponds to the line $r_1\cos(k-\theta)=1$, which is the tangent to the circle $r_1=1$ at $\theta=k$. Similarly, $\theta-\sigma=k$ corresponds to the same tangent. Along any such tangent, r_1 decreases from ∞ to 0 and then increases again to ∞, and the complete tangent therefore corresponds to two characteristics in the (σ,θ)-plane lying between $\sigma=0$ and $\sigma=\frac{1}{2}\pi$. Along one part of the tangent θ increases from $r_1=\infty$ to $r_1=1$; along the other part θ decreases from $r_1=\infty$ to $r_1=1$. These correspond to the characteristics in the (σ,θ)-plane along which θ increases and decreases, respectively, as σ decreases from $\frac{1}{2}\pi$ to 0, and therefore to the characteristics $\theta+\sigma=k$ and $\theta-\sigma=k$ respectively. Thus the characteristics $\theta+\sigma=$ *constant* correspond to those parts of the tangents to $r_1=1$ on which θ increases as the circle is approached, and the characteristics $\theta-\sigma=$ *constant* to those parts of the tangents on which θ decreases as the circle is approached. The correspondence is illustrated in fig. 7.3.

The mean surface of the wing in fig. 7.1 (*a*) corresponds, in the (x_1, y_1)-plane, to the part of the axis $y_1 = 0$ between

$$x_1 = -\sec\sigma_1 = -B\cot\chi_1 \quad \text{and} \quad x_1 = \sec\sigma_0 = B\cot\chi_0,$$

as in fig. 7.1 (*b*), and the above discussion shows that the regions 1, 2, 3, 4, in the (σ, θ)-plane (fig. 7.2) correspond to the regions 1, 2, 3, 4, in the (x_1, y_1)-plane (fig. 7.1 (*b*)), bounded by arcs of the circle $r_1 = 1$ (corresponding to $\sigma = 0$), by $y_1 = 0$ (corresponding to $\theta = 0$ and $\theta = \pi$), and by the tangents from the ends of the portion

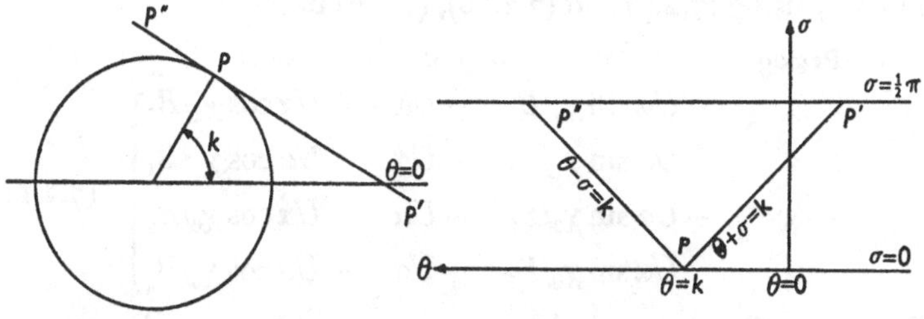

Fig. 7.3. The (x_1, y_1)-plane and the (σ, θ)-plane, showing the correspondence between the characteristic lines in the two planes.

of $y_1 = 0$ corresponding to the mean wing surface. Since the (x_1, y_1)-plane is effectively a cross-section of the flow-field by a plane $z = B$, and all cross-sections must be similar in conical fields, the regions of physical space corresponding to regions 1, 2, 3, 4, in the (x_1, y_1)-plane are easily visualized. They are bounded by the Mach cone of the vertex (corresponding to $r_1 = 1$), by $y = 0$ (corresponding to $y_1 = 0$), and by tangent planes to the Mach cone of the vertex which pass through the leading edges of the mean wing surface (corresponding to the tangents to $r_1 = 1$). These tangent planes are the envelopes of characteristic cones with their vertices on the leading edges, and hence are characteristic surfaces for the linearized equations; together with the appropriate portions of the Mach cone of the vertex, they bound the influence domain of the wing.

The values of u, v, w found above can be put into a more familiar form. Since

$$\sec\sigma_0 = B\cot\chi_0 = \cot\mu\cot\chi_0, \tag{7.2.7}$$

it follows that

$$\begin{aligned}\tan^2\sigma_0 &= \cot^2\mu\cot^2\chi_0 - 1 \\ &= \{(M^2-1)\cos^2\chi_0 - \sin^2\chi_0\}/\sin^2\chi_0,\end{aligned} \tag{7.2.8}$$

and hence
$$\left.\begin{aligned}\cot\sigma_0&=\frac{\sin\chi_0}{\sqrt{(M^2\cos^2\chi_0-1)}}=\frac{\sin\chi_0}{B_0},\quad\text{say,}\\ \operatorname{cosec}\sigma_0&=\frac{\cos\chi_0\sqrt{(M^2-1)}}{\sqrt{(M^2\cos^2\chi_0-1)}}=\frac{B\cos\chi_0}{B_0}.\end{aligned}\right\}\quad(7.2.9)$$

Similarly
$$\left.\begin{aligned}\cot\sigma_1&=\frac{\sin\chi_1}{\sqrt{(M^2\cos^2\chi_1-1)}}=\frac{\sin\chi_1}{B_1},\\ \operatorname{cosec}\sigma_1&=\frac{\cos\chi_1\sqrt{(M^2-1)}}{\sqrt{(M^2\cos^2\chi_1-1)}}=\frac{B\cos\chi_1}{B_1}.\end{aligned}\right\}\quad(7.2.10)$$

Thus by using (7.2.9) and (7.2.10), (7.2.6) becomes

Region	u	v	w
1	$-U\alpha\sin\chi_1/B_1$	$-U\alpha$	$-U\alpha\cos\chi_1/B_1$
2	$U\alpha\sin\chi_1/B_1$	$-U\alpha$	$U\alpha\cos\chi_1/B_1$
3	$-U\alpha\sin\chi_0/B_0$	$-U\alpha$	$U\alpha\cos\chi_0/B_0$
4	$U\alpha\sin\chi_0/B_0$	$-U\alpha$	$-U\alpha\cos\chi_0/B_0$

(7.2.11)

These are the same velocity components as for two-dimensional plane wings yawed through angles χ_0, $-\chi_1$, since the effect of the vertex of the wing is not felt outside the Mach cone of the field.

If one edge of the wing is outside the Mach cone, and the other is inside, then either regions 1 and 2, or regions 3 and 4 are absent, and the results for the remaining pair are unaltered, with $\mathbf{v}=0$ elsewhere for $r_1\geqslant 1$. If both edges are inside the Mach cone, then $\mathbf{v}=0$ everywhere for $r_1\geqslant 1$.

Similar results can be obtained when v is a variable quantity on the cuts in the (σ,θ)-plane and hence on the wing (see end of this section), and also in the cases when the wing is not plane, and the cuts in the (σ,θ)-plane are not necessarily on $\theta=0$ and $\theta=\pi$.

A slightly different case arises, however, when both edges are outside the Mach cone of the vertex, and on the same side of it. If the leading and trailing edges are swept back at angles χ_0, χ_1, $(\chi_0<\chi_1<\frac{1}{2}\pi-\mu)$, then the mean wing surface corresponds, in the (σ,θ)-plane, to the portion of $\theta=0$ for which $\sigma_1\leqslant\sigma\leqslant\sigma_0$, where σ_0, σ_1 are the same as before, in (7.2.1). The physical planes and the (σ,θ)-plane are shown in fig. 7.4 (*a*), (*b*), (*c*). Except in the regions 3, 4, 5, 6 in fig. 7.4 (*c*), in which the characteristic lines shown are parts of $\sigma\pm\theta=\sigma_0$, $\sigma\pm\theta=\sigma_1$, $\mathbf{v}$ is zero as before. The regions 5 and 6 are separated by $\theta=0$, $0<\sigma<\sigma_1$, and on this boundary the normal velocity and pressure must be continuous, i.e. v and w must be continuous.

The velocity components in regions 3 and 4 are the same as before. In passing into region 5 from region 3, $\mathbf{F}(\theta-\sigma)$ does not change; let $f_2(\theta+\sigma)$ increase by K as $\theta+\sigma$ decreases through σ_1, so v

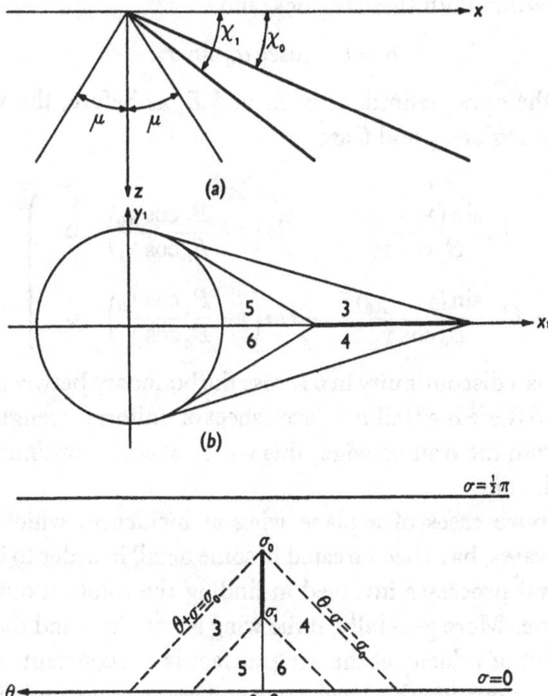

Fig. 7.4. (*a*) Wing leading and trailing edges in the (z, x)-plane, (*b*) trace of the wing in the (x_1, y_1)-plane, and (*c*) the corresponding (σ, θ)-plane, when both edges are outside and on the same side of the Mach cone of the vertex.

becomes $-U\alpha+K$ throughout region 5. By the same argument as before, u increases by $K\cot\sigma_1$, and w by $-K\operatorname{cosec}\sigma_1/B$. Hence, in region 5,

$$\left.\begin{aligned} u=-U\alpha\cot\sigma_0+K\cot\sigma_1, \quad v=-U\alpha+K, \\ w=(U\alpha\operatorname{cosec}\sigma_0-K\operatorname{cosec}\sigma_1)/B. \end{aligned}\right\} \qquad (7.2.12)$$

In passing into region 6 from region 4, $\mathbf{f}(\theta+\sigma)$ does not change, and $F_2(\theta-\sigma)$ must increase by K in order that v should be continuous on the boundary between regions 5 and 6. In region 6, therefore,

$$\left.\begin{aligned} u=U\alpha\cot\sigma_0-K\cot\sigma_1, \quad v=-U\alpha+K,\\ w=-(U\alpha\operatorname{cosec}\sigma_0-K\operatorname{cosec}\sigma_1)/B.\end{aligned}\right\} \tag{7.2.13}$$

Since w must be continuous on the boundary between regions 5 and 6, $w=0$ in both these regions, and

$$K=U\alpha\operatorname{cosec}\sigma_0\sin\sigma_1. \tag{7.2.14}$$

With the same definitions of B_0 and B_1 as before, the values of u, v, w, in regions 5 and 6 are

$$\left.\begin{array}{cccc} \text{Region} & u & v & w \\ 5 & U\alpha\dfrac{\sin(\chi_1-\chi_0)}{B_0\cos\chi_1} & -U\alpha\left(1-\dfrac{B_1\cos\chi_0}{B_0\cos\chi_1}\right) & 0 \\ 6 & -U\alpha\dfrac{\sin(\chi_1-\chi_0)}{B_0\cos\chi_1} & -U\alpha\left(1-\dfrac{B_1\cos\chi_0}{B_0\cos\chi_1}\right) & 0 \end{array}\right\} \tag{7.2.15}$$

There is a discontinuity in u across the boundary between regions 5 and 6, so there is a trailing vortex sheet of uniform strength downstream from the trailing edge; this vortex sheet is continued inside the Mach cone.

The above cases of a plane wing at incidence, which are the simplest cases, have been treated in some detail in order to illustrate the general processes involved in finding the solution outside the Mach cone. More generally, if the wing is not plane and the normal component of velocity at the wing surface is not constant, as above, the solution outside the Mach cone still assumes a simple form. In order that the problem should be of the conical type, the slope of the wing surfaces must be constant along lines passing through the wing vertex, so that the slopes are functions of σ only. Then, if the mean wing surface is the plane $y=0$ as before, the boundary conditions on the mean surface take the form

$$v=U\eta(\sigma), \tag{7.2.16}$$

where $\eta(\sigma)$ is the slope of the relevant wing surface.

As an example, suppose that the slope of the wing surface in region 3 of figs. 7.1 (*b*) and 7.2 is $\eta_0(\sigma)$ on $\theta=+0$, $0<\sigma<\sigma_0$. Then v in region 3 is given by

$$v=f_2(\theta+\sigma)=U\eta_0(\theta+\sigma). \tag{7.2.17}$$

This result shows that v is constant on characteristic lines $\theta+\sigma=$ *constant* in region 3 in the (σ,θ)-plane, and hence that v is constant on the characteristic lines (the tangents to $r_1=1$) in the (x_1,y_1)-plane. The other velocity components can be calculated from (7.1.9). Since

$$v=\int^{\theta+\sigma} f(\lambda)\sin\lambda\,d\lambda \tag{7.2.18}$$

in region 3, and $v=0$ on $\theta=0$, $\sigma_0<\sigma<\frac{1}{2}\pi$, it follows that

$$v=U\int_{\frac{1}{2}\pi}^{\theta+\sigma} d\eta_0(\lambda) \tag{7.2.19}$$

in region 3, where the integral is a Stieltjes integral, and $\eta_0(\lambda)=0$ in $\lambda>\sigma_0$. Hence, in this region,

$$\left.\begin{aligned} u&=U\int_{\frac{1}{2}\pi}^{\theta+\sigma}\cot\lambda\,d\eta_0(\lambda),\\ w&=-\frac{U}{B}\int_{\frac{1}{2}\pi}^{\theta+\sigma}\operatorname{cosec}\lambda\,d\eta_0(\lambda).\end{aligned}\right\} \tag{7.2.20}$$

These expressions for the velocity components do not require $\eta_0(\sigma)$ to be a continuous function of σ, the discontinuities being taken into account by the Stieltjes form of the integrals.

Similar expressions are easily written down for the regions 1, 2 and 4 when the edges are on opposite sides of the Mach cone, or for the regions 4, 5 and 6 when the edges are on the same side of the Mach cone.

7.3. Flow inside the Mach cone of the vertex

For $r_1\leqslant 1$, $\mathbf{v}$ is the real part of a function of ζ, as in (7.1.16). From the results of §7.2, $\mathbf{v}$ is known on $r_1=1$ in the (x_1,y_1)-plane, and therefore on the circle $|\zeta|=1$ in the ζ-plane. Also the normal velocity is known on any cuts, inside $|\zeta|=1$, which are the traces of the mean surfaces of the body under consideration. In the cases considered in §7.2, the mean wing surface corresponds to part of $y_1=0$, and v is known there. The part of the wing inside the Mach cone of the vertex corresponds to the whole or part of the axis $y_1=0$ between $x_1=\pm 1$, and therefore to the whole or part of the real axis in the ζ-plane inside the circle $|\zeta|=1$; along this part of the real axis in the ζ-plane, $v=-U\alpha$. Hence for $|\zeta|\leqslant 1$, the real part of $g_2(\zeta)$

(cf. (7.1.14)) is known on $|\zeta|=1$ and on any cut there may be inside it, and $g_2(\zeta)$ can be found at all points inside $|\zeta|=1$ so that the velocity satisfies the conditions of §4.5. When $g_2(\zeta)$ has been determined, $g_1(\zeta)$ and $g_3(\zeta)$ can be found from the relation (7.1.17) together with the values of u and w at any one point on $r_1=1$.

In the case of flat wings at incidence, and in many other cases, it is easier to work in a t-plane, defined by the transformation

$$1/t=\tfrac{1}{2}(\zeta+1/\zeta), \tag{7.3.1}$$

rather than in the ζ-plane directly. From (7.1.13),

$$t=r_1/\{\cos\theta-i\sqrt{(1-r_1^2)}\sin\theta\}, \tag{7.3.2}$$

which gives the direct relation between t and (r_1,θ) in the (x_1,y_1)-plane.

The circle $r_1=1$ and its interior correspond with the circle $|\zeta|=1$ and its interior; the circles are transformed, in the t-plane, into the upper and lower sides of the parts of the real axis for which $|t|\geqslant 1$; their interiors are represented on the whole t-plane with cuts along the portions $|t|\geqslant 1$ of the real axis, the upper halves $(0<\theta<\pi)$ being represented on the upper half-plane $(\mathfrak{I}(t)>0)$ and the lower halves $(-\pi<\theta<0)$ on the lower half-plane $(\mathfrak{I}(t)<0)$. The axis $x_1=0$ and the imaginary axis in the ζ-plane, inside the unit circle in each case, correspond with one another, and are transformed into the imaginary axis in the t-plane. This correspondence in shown in fig. 7.5.

The point $(x_1,0)$ on the x_1-axis inside the unit circle in the (x_1,y_1)-plane corresponds with the point $t=x_1$ on the real axis in the t-plane, from (7.3.2). For a point on the unit circle in the (x_1,y_1)-plane, or in the ζ-plane, $t=\sec\theta$; at $\theta=\pm\sigma$, $t=\sec\sigma\pm 0i$ and $\sec\sigma$ is the abscissa, in the (x_1,y_1)-plane, of the intersection with the x_1-axis of the tangents to the unit circle at $\theta=\pm\sigma$. Thus the point $t=x_1$ on the real axis in the t-plane corresponds with the point $(x_1,0)$ if $|x_1|<1$, and with the points of contact of the tangents to the unit circle from $(x_1,0)$ if $|x_1|>1$. This correspondence is illustrated in fig. 7.6.

The results of §7.2 show that the value of $\mathbf{v}$ at A' in fig. 7.6 is equal to the value of $\mathbf{v}$ at A; hence it follows that *the boundary condition on the real axis in the t-plane is formally the same as that on the x_1-axis in the (x_1,y_1)-plane.*

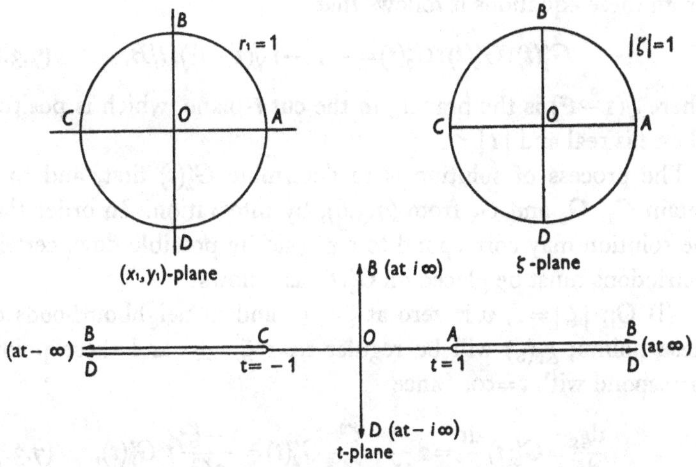

Fig. 7.5. The correspondence between the (x_1, y_1)-plane, the ζ-plane, and the t-plane, for the flow inside the Mach cone of the vertex.

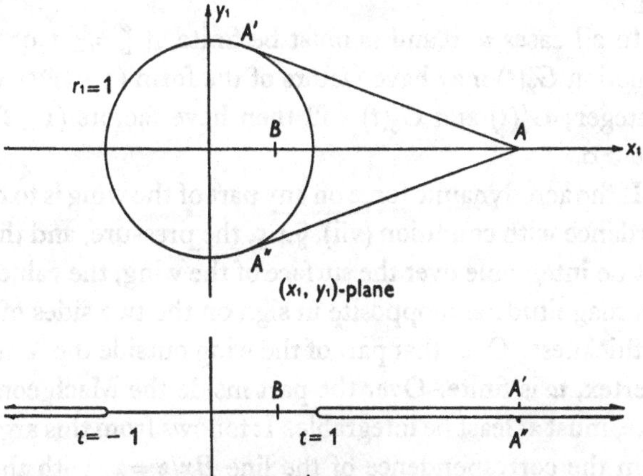

Fig. 7.6. The correspondence between the trace of the Mach cone of the vertex in the (x_1, y_1)-plane, the x_1-axis, and the real axis in the t-plane.

The velocity components u, v and w are the real parts of functions of t,

$$u = \Re G_1(t), \quad v = \Re G_2(t), \quad w = \Re G_3(t), \qquad (7.3.3)$$

and, from (7.1.20),

$$G_1'(t) : G_2'(t) : G_3'(t) = -(h^2+1) : i(h^2-1) : 2h/B, \qquad (7.3.4)$$

where now $\quad \zeta = h(t) \quad$ and $\quad 1/t = \tfrac{1}{2}(\zeta + 1/\zeta). \qquad (7.3.5)$

From these equations it follows that

$$G_1'(t):G_2'(t):G_3'(t) = -1:-i\sqrt{(1-t^2)}:t/B, \qquad (7.3.6)$$

where $\sqrt{(1-t^2)}$ is the branch, in the cut t-plane, which is positive when t is real and $|t|<1$.

The process of solution is to determine $G_2'(t)$ first, and then obtain G_1, G_2 and G_3 from (7.3.6), by integration. In order that the solution may correspond to a physically possible flow, certain restrictions must be placed on $G_2'(t)$, as follows:

(i) On $|\zeta|=1$, v is zero at $\zeta=\pm i$ and in neighbourhoods of these points; $g_2(\zeta)$ will be regular near $\zeta=\pm i$ and these points correspond with $t=\infty$. Since

$$\frac{dg_2}{d\zeta} = G_2'(t)\frac{dt}{d\zeta} = 2\frac{1-\zeta^2}{(1+\zeta^2)^2}G_2'(t) = \frac{1-\zeta^2}{2\zeta^2}t^2G_2'(t), \qquad (7.3.7)$$

$G_2'(t)$ must be $O(t^{-2})$ as $|t|\to\infty$ in order that $g_2'(\zeta)$ should be finite at $\zeta=\pm i$.

(ii) In all cases u, v and w must be finite at $\zeta=\pm 1$ or $t=\pm 1$. The function $G_2'(t)$ may have factors of the form $(1\pm t)^{n+\frac{1}{2}}$ where n is an integer; $G_1'(t)$ and $G_3'(t)$ will then have factors $(1\pm t)^n$, so n must be $\geqslant 0$.

(iii) If the aerodynamic force on any part of the wing is to be finite in accordance with condition (vii), § 4.5, the pressure, and therefore w, must be integrable over the surface of the wing, the values being equal in magnitude and opposite in sign on the two sides of a wing of zero thickness. Over that part of the wing outside the Mach cone of its vertex, w is finite. Over the part inside the Mach cone, w, if not finite, must at least be integrable. It follows from this argument, and from the correspondence of the line $Bx/z=x_1$ with the point $t=x_1$, that $w(=\Re G_3(t))$ must be integrable along the real axis in the t-plane between the extreme limits of x_1 on the wing. If t_1 is one of the limits of x_1 on the wing, then $G_2'(t)$, and therefore also $G_3'(t)$, may have factors of the form $(t-t_1)^m$ and $(t-t_1)^{n+\frac{1}{2}}$, where m and n are integers. The preceding argument shows that m must be $\geqslant -1$, and n must be $\geqslant -2$. These conditions are also sufficient to ensure that the suction forces on subsonic edges are finite.

In other problems, where, for example, the wing is not flat and the t-plane is not appropriate for the solution, similar considerations to the above have to be used in order to restrict the solution to physically possible flows.

In order to facilitate a comparison of the results of §7.2 for $r_1 \geqslant 1$ with those obtained by the methods of the present section for $r_1 \leqslant 1$, the values of u, v and w given in (7.2.11) and (7.2.15) are written below in the notation of the t-plane. Only the regions in $y_1 > 0$ need be considered since the values of u, v and w in $y_1 < 0$ are obtained from those in $y_1 > 0$ from the fact that u and w are odd functions of y_1 and v is an even function of y_1 for the cases considered.

Let

$$\left.\begin{aligned} t_0 &= \sec\sigma_0 = B\cot\chi_0 > 1,\\ t_1 &= \sec\sigma_1 = B\cot\chi_1 > 1,\end{aligned}\right\} \tag{7.3.8}$$

then

$$\left.\begin{aligned} \cot\sigma_0 &= 1/\surd(t_0^2-1), \quad \operatorname{cosec}\sigma_0 = t_0/\surd(t_0^2-1),\\ \cot\sigma_1 &= 1/\surd(t_1^2-1), \quad \operatorname{cosec}\sigma_1 = t_1/\surd(t_1^2-1).\end{aligned}\right\} \tag{7.3.9}$$

When both edges are outside the Mach cone and on opposite sides of it, (7.2.11) becomes

Region	u	v	w
2	$U\alpha/\surd(t_1^2-1)$	$-U\alpha$	$U\alpha t_1/B\surd(t_1^2-1)$
3	$-U\alpha/\surd(t_0^2-1)$	$-U\alpha$	$U\alpha t_0/B\surd(t_0^2-1)$

(7.3.10)

and when both edges are outside the Mach cone, and on the same side of it, the values in region 3 are as above, and the values in region 5 are

$$u = \frac{U\alpha(t_0-t_1)}{t_1\surd(t_0^2-1)}, \quad v = -U\alpha\left(1-\frac{t_0\surd(t_1^2-1)}{t_1\surd(t_0^2-1)}\right), \quad w = 0. \tag{7.3.11}$$

The result most often required is the pressure on the wing surface, which is given from the linearized Bernoulli equation, and its value may be taken on the mean surface. In §7.5, the value of w, which gives the pressure, is given on the upper side of the mean surface, $y_1 = +0$, for plane wings at incidence. The value on the lower side, $y = -0$, is equal in magnitude, and opposite in sign.

7.4. The flow inside the Mach cone of the vertex: the symmetrical problem for nearly plane wings

If the mean plane of the wing is $y = 0$, then the trace of the wing in the (x_1, y_1)-plane is part of the x_1-axis. Let the sweeps of the edges be χ_0 and χ_1; then, if both edges are leading edges, the trace is that part of the x_1-axis for which

$$-B\cot\chi_1 < x_1 < B\cot\chi_0, \tag{7.4.1}$$

and if one edge (χ_0) is a leading edge, and the other (χ_1) is a trailing edge, and the whole wing lies in $x>0$, the trace is that part of the x_1-axis for which

$$B\cot\chi_1 < x_1 < B\cot\chi_0. \tag{7.4.2}$$

The boundary condition on the upper wing surface ($y=+0$) is

$$\left.\begin{aligned} v &= U\eta(x_1) \quad \text{on the wing trace in } y_1=+0,\\ v &= 0 \qquad\quad\ \ \text{everywhere else in } y_1=0, \end{aligned}\right\} \tag{7.4.3}$$

where $\eta(x_1)$ is the slope of the wing surface for $y=+0$.

In the t-plane, let $t=t_0$, $t=t_1$ denote the limits of the trace of the wing; then the boundary condition for the flow in $\mathfrak{I}(t)>0$ is

$$\left.\begin{aligned} v &= U\eta(t) \quad \text{for } t_1<t<t_0\\ v &= 0 \qquad\quad \text{for } t<t_1,\ t>t_0 \end{aligned}\right\} \text{ on } \mathfrak{I}(t)=+0. \tag{7.4.4}$$

Now from (7.3.3), v is the real part of an analytic function, $G_2(t)$, of t in the upper half-plane, and, from (7.4.4), the real part of this function is known on the real axis. Hence $G_2(t)$ can be written down immediately in the form

$$G_2(t) = \frac{U}{\pi \mathrm{i}}\int_{t_1}^{t_0}\frac{\eta(t')}{t'-t}\,\mathrm{d}t', \tag{7.4.5}$$

and v is the real part of this integral. Expressions for u and w, being the real parts of $G_1(t)$ and $G_3(t)$, can be written down immediately by differentiating under the integral sign in (7.4.5), using (7.3.6), and integrating again. However, it appears to be simpler in practice to evaluate (7.4.5) with a given $\eta(t')$ *before* differentiating, and then to use (7.3.6), thus avoiding the occurrence of repeated integrals. This procedure is illustrated in connexion with the anti-symmetrical problem in the next section.

7.5 Flow inside the Mach cone of the vertex: the anti-symmetrical problem for plane wings

(i) *Both edges outside the Mach cone on opposite sides.* For this case (fig. 7.1), for $y_1>0$, u, v and w vanish for $r_1 \geqslant 1$ except in regions 2 and 3, where their values are given by (7.2.6), (7.2.11) or

(7.3.10). Hence, on the real axis in the t-plane,

$$\left.\begin{aligned} v&=\Re G_2(t)=-U\alpha \quad \text{for} \quad -t_1<t<t_0, \\ v&=0 \quad \text{for} \quad t>t_0,\ t<-t_1. \end{aligned}\right\} \tag{7.5.1}$$

The solution for $G_2(t)$ can be written down immediately from (7.4.5), with $\eta(t)=-U\alpha$, and $-t_1$ instead of t_1:

$$G_2(t)=-\frac{U\alpha}{\pi i}\log\left(\frac{t-t_0}{t+t_1}\right), \tag{7.5.2}$$

with $\arg(t-t_0)=\arg(t+t_1)=0$ for t real and $>t_0$. Then

$$G_2'(t)=\frac{U\alpha(t_0+t_1)}{\pi i(t-t_0)(t+t_1)} \tag{7.5.3}$$

(conditions (i), (ii) and (iii) of §7.3 are satisfied), and from (7.3.6), on integrating and taking $G_3=0$ at $t=\infty$,

$$G_3(t)=\frac{U\alpha}{B\pi}\left\{\frac{t_0}{\sqrt{(t_0^2-1)}}\cos^{-1}\left(\frac{1-t_0t}{t_0-t}\right)+\frac{t_1}{\sqrt{(t_1^2-1)}}\cos^{-1}\left(\frac{1+t_1t}{t_1+t}\right)\right\} \tag{7.5.4}$$

in the upper half-plane, where the inverse cosines are between 0 and π for real arguments. The velocity component w is found by taking the real part of this function.†

The value of w on $y=+0$ for $|x_1|<1$ is, from (7.5.4),

$$w=\frac{U\alpha}{\pi}\left\{\frac{\cos\chi_0}{B_0}\cos^{-1}\left(\frac{1-t_0x_1}{t_0-x_1}\right)+\frac{\cos\chi_1}{B_1}\cos^{-1}\left(\frac{1+t_1x_1}{t_1+x_1}\right)\right\} \tag{7.5.5}$$

(cf. (7.2.11) for the value of w on $y=+0$ for $|x_1|>1$).

(ii) *Both edges outside the Mach cone, on the same side.* For this case (fig. 7.4), for $y_1>0$, u, v and w vanish for $r_1\geqslant 1$ except in regions 3 and 5. Their values in region 3 are given in (7.3.10), and in region 5 in (7.3.11), (or in (7.2.11) and (7.2.15)). Hence, on the real axis in the t-plane,

$$\left.\begin{aligned} v&=0 \quad \text{for } t<-1,\ t>t_0, \\ v&=-U\alpha \quad \text{for } t_1<t<t_0, \\ v&=-U\alpha\left(1-\frac{t_0\sqrt{(t_1^2-1)}}{t_1\sqrt{(t_0^2-1)}}\right) \quad \text{for } 1<t<t_1, \end{aligned}\right\} \tag{7.5.6}$$

so that $\Re G_2'(t)=0$ on the real axis for $|t|>1$, $t\neq t_0$ or t_1.

† Complete expressions for the velocity components are given in the paper by Goldstein and Ward (1950).

The pressure, and therefore w also, are odd functions of y_1, and are continuous across the axis $y_1=0$ inside $r_1=1$. Hence $w=0$ on $y_1=0$, $|x_1|<1$, i.e. $w=0$ on the real axis in the t-plane for $|t|<1$. It follows that $\Re G_3'(t)=0$ and, from (7.3.6), that $\Im G_2'(t)=0$ on the real axis in the t-plane for $|t|<1$. Hence, on the real axis

$$\left.\begin{aligned}\Re G_2'(t)&=0 \quad \text{for } |t|>1,\ t\neq t_0 \text{ or } t_1,\\ \Im G_2'(t)&=0 \quad \text{for } |t|<1,\end{aligned}\right\} \tag{7.5.7}$$

and $G_2(t)$ has finite discontinuities at t_0 and t_1. Therefore $G_2'(t)$ must have the form

$$G_2'(t)=\frac{(1+t)^{a+\frac{1}{2}}(1-t)^{b+\frac{1}{2}}}{(t-t_0)(t-t_1)}G(t), \tag{7.5.8}$$

where a and b are integers, and $G(t)$ is a rational function of t, real when t is real. The residues at the poles t_0 and t_1, multiplied by πi, must be equal to the discontinuities in $G_2(t)$ at those points along the real axis.

If $G(t)=O(t^{-m})$ as $t\to\infty$, then, from condition (i) of §7.3, $m\geqslant a+b+1$; from condition (ii), $a\geqslant 0$, $b\geqslant 0$. Hence $m\geqslant 1$, and $G(t)$ has a singularity which is at least a pole; if it is a pole, $a=b=0$, $m=1$, and $G(t)$ is of the form *constant*$/(t-t_2)$.

It was found in §7.2 that the boundary between regions 5 and 6 contains a vortex sheet of constant strength. The vortex lines in linearized theory must be parallel to the z-axis, so the linearized vortex sheet must be of uniform strength, bounded by the z-axis. Hence u must have a finite discontinuity at $t=0$, so $G(t)$ must have a pole at $t=0$ and be equal to A_1/t, where A_1 is a constant. That the singularity in $G(t)$ must be at $t=0$, and cannot be a higher singularity than a pole also follows from the fact that the pressure, and therefore w, must be continuous along the real axis in the t-plane except at t_0 and t_1. The value of A_1 is found by equating the residue of $G_2'(t)$ at $t=t_0$ to $iU\alpha/\pi$. Hence

$$G_2'(t)=-\frac{U\alpha t_0(t_0-t_1)}{\pi\sqrt{(t_0^2-1)}}\frac{\sqrt{(1-t^2)}}{t(t-t_0)(t-t_1)}, \tag{7.5.9}$$

and from (7.3.6), on integrating for the upper half-plane,

$$G_3(t)=\frac{U\alpha}{\pi i B}\frac{t_0}{\sqrt{(t_0^2-1)}}\log\left(\frac{t-t_0}{t-t_1}\right). \tag{7.5.10}$$

The vortex sheet is of constant strength

$$\frac{2U\alpha(t_0-t_1)}{t_0\sqrt{(t_0^2-1)}}. \tag{7.5.11}$$

There is a logarithmic infinity in v at $t=0$, i.e. at the edge of the linearized vortex sheet.

(iii) *One edge outside and one edge inside the Mach cone.* Let the edges be swept back at angles χ_0, χ_1, where $\chi_0<\frac{1}{2}\pi-\mu<\chi_1<\frac{1}{2}\pi+\mu$. There are two cases which must be distinguished, depending on whether or not the wing extends through the z-axis. In the case when it does, $\chi_1>\frac{1}{2}\pi$, and the edge inside the Mach cone is a subsonic leading edge. In the case when it does not, $\chi_1<\frac{1}{2}\pi$, and the edge is a subsonic trailing edge at which the Kutta-Joukowski

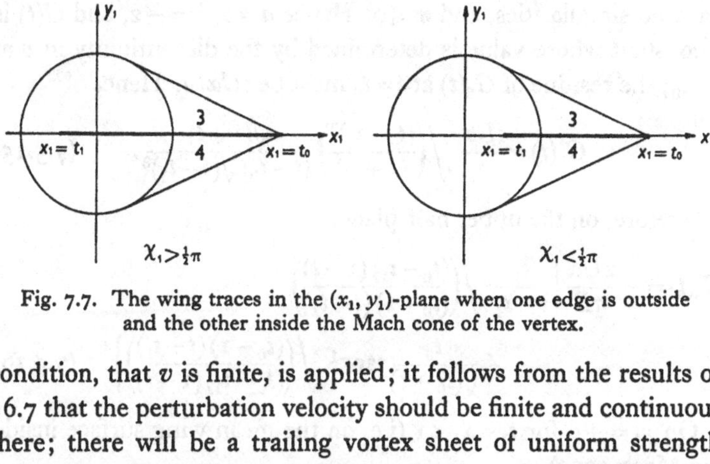

Fig. 7.7. The wing traces in the (x_1, y_1)-plane when one edge is outside and the other inside the Mach cone of the vertex.

condition, that v is finite, is applied; it follows from the results of §6.7 that the perturbation velocity should be finite and continuous there; there will be a trailing vortex sheet of uniform strength downstream from this edge, with its other boundary on the z-axis. The (x_1, y_1)-planes for these two cases are shown in fig. 7.7. In both cases the velocities in region 3 are those given in (7.2.11) or (7.3.10).

The boundary conditions in the t-plane are the same in both cases; the differences in the solutions come from differences in the singularities. By the same arguments as in (ii) above, on the real axis in the t-plane,

$$\left.\begin{array}{ll} v=-U\alpha & \text{for } t_1<t<t_0, \\ v=0 & \text{for } t<-1,\ t>t_0, \\ w=0 & \text{for } -1<t<t_1, \end{array}\right\} \tag{7.5.12}$$

where $t_0 = B\cot\chi_0$, $t_1 = B\cot\chi_1$ as usual. Hence, on the real axis,

$$\left.\begin{aligned}\Re G_2'(t)&=0 \quad \text{for } t<-1,\ t>t_1,\ t\neq t_0,\\ \Im G_2'(t)&=0 \quad \text{for } -1<t<t_1,\end{aligned}\right\} \tag{7.5.13}$$

and $G_2'(t)$ is of the form

$$G_2'(t)=\frac{i(1+t)^{a+\frac{1}{2}}(t-t_1)^{b+\frac{1}{2}}}{t-t_0}G(t), \tag{7.5.14}$$

where a and b are integers and $G(t)$ is a rational function of t, real when t is real. From conditions (ii) and (iii) of §7.3, $a \geq 0$, $b \geq -2$; from condition (i), if $G(t)=O(t^{-m})$ as $t\to\infty$, $m \geq a+b+2$.

In the first case, when $\chi_1 > \frac{1}{2}\pi$, t_1 is negative, the wing passes through the z-axis, and there is no trailing vortex sheet. $G(t)$ will have no singularities, and $m \leq 0$. Hence $a=0$, $b=-2$, and $G(t)$ is a constant whose value is determined by the discontinuity in v at $t=t_0$; the residue of $G_2(t)$ at $t=t_0$ must be $iU\alpha/\pi$. Hence

$$G_2'(t)=\frac{iU\alpha}{\pi}\sqrt{\left(\frac{(t_0-t_1)^3}{(1+t_0)}\right)}\frac{\sqrt{(1+t)}}{(t-t_0)\sqrt{(t-t_1)^3}}. \tag{7.5.15}$$

Therefore, on the upper half-plane,

$$\begin{aligned}G_3(t)=-\frac{2U\alpha}{\pi B}\Bigg\{&\frac{t_1}{1-t_1}\sqrt{\left(\frac{(t_0-t_1)(1-t)}{(t_0+1)(t-t_1)}\right)}\\ &-\frac{t_0}{\sqrt{(t_0^2-1)}}\tan^{-1}\sqrt{\left(\frac{(t_0-1)(t-t_1)}{(t_0-t_1)(1-t)}\right)}\Bigg\}.\end{aligned} \tag{7.5.16}$$

On $y_1=+0$ for $t_1<x_1<1$ (i.e. on the mean wing surface inside the Mach cone),

$$\begin{aligned}w=\frac{2U\alpha}{\pi}\Bigg\{&\frac{-\cot\chi_1}{1-t_1}\sqrt{\left(\frac{(t_0-t_1)(1-x_1)}{(t_0+1)(x_1-t_1)}\right)}\\ &+\frac{\cos\chi_0}{B_0}\tan^{-1}\sqrt{\left(\frac{(t_0-1)(x_1-t_1)}{(t_0-t_1)(1-x_1)}\right)}\Bigg\},\end{aligned} \tag{7.5.17}$$

$\cot\chi_1$ and t_1 being negative here.

In the second case, when $\chi_1 < \frac{1}{2}\pi$, t_1 is positive, and there is a trailing vortex sheet. $G_2'(t)$ is still given by (7.5.14), but the Kutta-Joukowski condition that v shall be finite at $t=t_1$, requires that $b \geq -1$. Hence $m \geq 1$, and $G(t)$ has a singularity. By the same argument as above, in (i), the singularity in $G(t)$ must be a pole at

$t = 0$, and $a = 0$, $b = -1$, $G(t) = constant/t$. The constant is determined from the discontinuity in v at $t = t_0$ as before, and

$$G_2'(t) = \frac{iU\alpha}{\pi} \frac{t_0\sqrt{(t_0-t_1)}}{\sqrt{(t_0+1)}} \frac{\sqrt{(1+t)}}{t(t-t_0)\sqrt{(t-t_1)}}. \tag{7.5.18}$$

Hence, on the upper half-plane,

$$G_3(t) = \frac{2U\alpha}{\pi B} \frac{t_0}{\sqrt{(t_0^2-1)}} \tan^{-1} \sqrt{\left(\frac{(t_0-1)(t-t_1)}{(t_0-t_1)(1-t)}\right)}. \tag{7.5.19}$$

On $y_1 = +0$, for $t_1 < x_1 < 1$,

$$w = \frac{2U\alpha \cos\chi_0}{\pi B_0} \tan^{-1} \sqrt{\left(\frac{(t_0-1)(x_1-t_1)}{(t_0-t_1)(1-x_1)}\right)}. \tag{7.5.20}$$

The vortex sheet is of constant strength

$$2U\alpha \sqrt{\left(\frac{t_0-t_1}{t_1(t_0+1)}\right)}. \tag{7.5.21}$$

(iv) *The tip of a swept-back wing.* In fig. 7.8, let O' be the tip of a wing whose vertex is at O, and let OO' be a straight leading edge swept back at an angle $\chi_0 < \frac{1}{2}\pi - \mu$. With O' as origin, this edge appears as a supersonic leading edge swept forward at an angle χ_0.

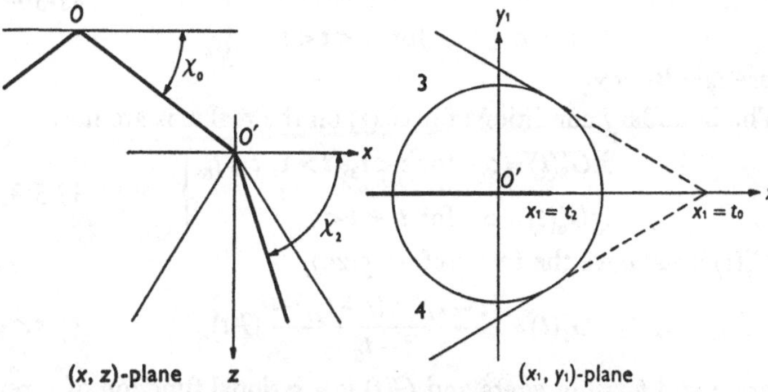

Fig. 7.8. Wing edges, and the trace of the wing in the (x_1, y_1)-plane, at a wing-tip.

Let the other edge through O' be swept back at an angle χ_2; then four cases arise,

$$0 < \chi_2 < \mu, \quad \mu < \chi_2 < \tfrac{1}{2}\pi, \quad \tfrac{1}{2}\pi < \chi_2 < \tfrac{1}{2}\pi + \mu, \quad \tfrac{1}{2}\pi + \mu < \chi_2 < \pi + \chi_0.$$

The problem is to find the velocity field inside the portion of the Mach cone of O' that is not intersected by the Mach cone of O.

The solution can be found in all cases by using the standard methods set out above; as an illustration the solutions for the second case (which is that shown in fig. 7.8) and the third case are set out below.

The velocity over the wing between the two Mach cones from O and O' is given in §7.2 for regions 3 and 4. In the (x_1, y_1)-plane with O' as origin, these results give u, v and w in region 3 of fig. 7.8, and, in particular, over its curved boundary. Hence, with t_0 defined as before, the value of v for region 3 gives its value on the upper side of the real axis in the t-plane for $t > t_0$ and $t < -1$, since the relevant portion of the unit circle in the (x_1, y_1)-plane transforms into these parts of the real axis in the t-plane. On that part of the unit circle in the (x_1, y_1)-plane which does not form the boundaries of regions 3 and 4, the perturbation velocity vanishes, so, in particular, $v=0$ there and on the real axis in the t-plane for $1 \leqslant t < t_0$. On the mean wing surface inside the Mach cone, $v = -U\alpha$, and on the rest of the x_1-axis inside the Mach cone, $w=0$ by the same reasoning as above. Thus the boundary conditions on the real axis in the t-plane are

$$\left.\begin{aligned} v &= -U\alpha \quad \text{for } t < t_2,\ t > t_0, \\ v &= 0 \quad\quad\ \text{for } 1 < t < t_0, \\ w &= 0 \quad\quad\ \text{for } t_2 < t < 1, \end{aligned}\right\} \tag{7.5.22}$$

where $t_2 = B \cot \chi_2$.

The boundary conditions for $G_2'(t)$ on the real axis are now

$$\left.\begin{aligned} \Re G_2'(t) &= 0 \quad \text{for } t < t_2,\ t > 1,\ t \neq t_0, \\ \Im G_2'(t) &= 0 \quad \text{for } t_2 < t < 1, \end{aligned}\right\} \tag{7.5.23}$$

so $G_2'(t)$ must have the form (cf. (7.5.20))

$$G_2'(t) = \frac{(1-t)^{a+\frac{1}{2}}(t-t_2)^{b+\frac{1}{2}}}{t-t_0} G(t), \tag{7.5.24}$$

where a and b are integers and $G(t)$ is a rational function of t, real when t is real.

The solution now proceeds in the same way as before. For the case when $\mu < \chi_2 < \frac{1}{2}\pi$, $t_2 > 0$, and there is no vortex sheet: on $y_1 = +0$, $-1 < x_1 < t_2$ (i.e. on the upper side of the mean wing surface inside the Mach cone of O')

$$w = \frac{2U\alpha}{\pi}\left\{\frac{\cot\chi_2}{1+t_2}\sqrt{\left(\frac{(t_0-t_2)(1+x_1)}{(t_0-1)(t_2-x_1)}\right)} + \frac{\cos\chi_0}{B_0}\tan^{-1}\sqrt{\left(\frac{(t_0+1)(t_2-x_1)}{(t_0-t_2)(1+x_1)}\right)}\right\}. \tag{7.5.25}$$

For the case when $\frac{1}{2}\pi < \chi_2 < \frac{1}{2}\pi + \mu$, $t_2 < 0$, and there is a vortex sheet: on $y_1 = +0$, $-1 < x_1 < t_2$ (i.e. on the upper side of the mean wing surface inside the Mach cone of O')

$$w = \frac{2U\alpha\cos\chi_0}{\pi B_0}\tan^{-1}\sqrt{\left(\frac{(t_0+1)(t_2-x_1)}{(t_0-t_2)(1+x_1)}\right)}, \qquad (7.5.26)$$

and the strength of the vortex sheet is

$$2U\alpha\sqrt{\left(\frac{t_0-t_2}{-t_2(t_0-1)}\right)}. \qquad (7.5.27)$$

Full expressions for $G_1(t)$, $G_2(t)$ and $G_3(t)$ on the upper half-plane are given in the paper by Goldstein and Ward referred to above.

(v) *Both edges inside the Mach cone.* Let the edges be swept back at angles χ_0, χ_1, where $\frac{1}{2}\pi - \mu < \chi_0 < \frac{1}{2}\pi$ and $\chi_0 < \chi_1 < \frac{1}{2}\pi + \mu$ (see fig. 7.9). Both edges are subsonic edges.

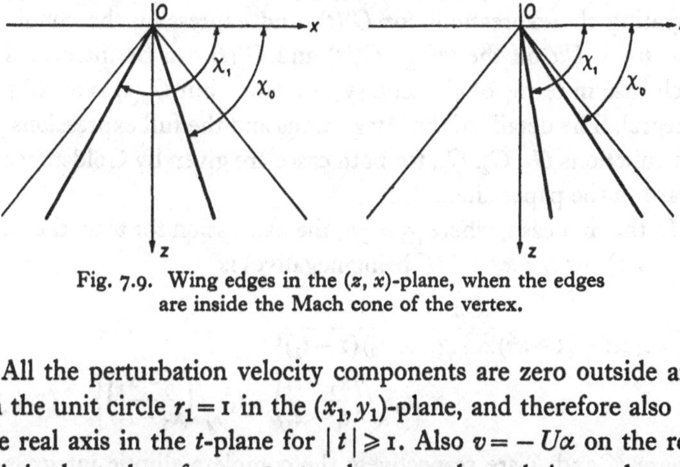

Fig. 7.9. Wing edges in the (z, x)-plane, when the edges are inside the Mach cone of the vertex.

All the perturbation velocity components are zero outside and on the unit circle $r_1 = 1$ in the (x_1, y_1)-plane, and therefore also on the real axis in the t-plane for $|t| \geqslant 1$. Also $v = -U\alpha$ on the real axis in the t-plane for $t_1 < t < t_0$, where t_0, t_1 have their usual meanings, and, by the same reasoning as before, $w = 0$ for $-1 < t < t_1$, $t_0 < t < 1$. Hence the boundary conditions for $G_2'(t)$ on the real axis are

$$\left.\begin{aligned} \Re G_2'(t) &= 0 \quad \text{for } t < -1,\ t_1 < t < t_0,\ t > 1,\\ \Im G_2'(t) &= 0 \quad \text{for } -1 < t < t_1,\ t_0 < t < 1, \end{aligned}\right\} \qquad (7.5.28)$$

so $G_2'(t)$ must have the form

$$G_2'(t) = (1-t)^{a+\frac{1}{2}}(1+t)^{b+\frac{1}{2}}(t-t_0)^{c+\frac{1}{2}}(t-t_1)^{d+\frac{1}{2}}\,G(t), \quad (7.5.29)$$

where a, b, c, d are integers and $G(t)$ is a rational function of t, real when t is real.

Again there are two cases, as shown in fig. 7.9, and the function $G(t)$ can be determined for each case by the same arguments as before. For the case when $\chi_1 < \frac{1}{2}\pi$, $t_1 < 0$, the wing extends through the z-axis, both edges are leading subsonic edges at which singularities are to be expected in linearized theory, and it is easily shown that

$$a = b = 0, \quad c = d = -2, \quad G(t) = \text{constant}. \qquad (7.5.30)$$

For the case when $\chi_1 < \frac{1}{2}\pi$, $t_1 > 0$, the wing does not extend through the z-axis, there is a trailing vortex sheet from the subsonic trailing edge at $t = t_1$ if the Kutta-Joukowski condition is applied to make v finite at $t = t_1$, and it is again easy to show that

$$a = b = 0, \quad c = -2, \quad d = -1, \quad G(t) = \text{constant}/t. \quad (7.5.31)$$

The constants in (7.5.30) and (7.5.31) have to be found by integrating the expressions for $G_2'(t)$, and expressing the condition that $v = -U\alpha$ on the wing. $G_1'(t)$ and $G_3'(t)$ can be integrated in each case in terms of elementary functions, but $G_2(t)$ is an elliptic integral. The details of the integrations and the full expressions for the functions G_1, G_2, G_3, for both cases are given by Goldstein and Ward in the paper cited.

In the first case, where $\chi_1 > \frac{1}{2}\pi$, the expression for w on the wing ($y = +0$) for $t_1 < x_1 < t_0$ (t_1 being negative) is

$$w = \frac{U\alpha}{B\{2E - (1-k^2)K\}\sqrt{\{(1+t_0)(1-t_1)\}}} \times \left\{t_0\sqrt{\left(\frac{x_1 - t_1}{t_0 - x_1}\right)} - t_1\sqrt{\left(\frac{t_0 - x_1}{x_1 - t_1}\right)}\right\}, \qquad (7.5.32)$$

where K and E are respectively the complete elliptic integrals of the first and second kind of modulus k, where

$$k^2 = \frac{(1-t_0)(1+t_1)}{(1+t_0)(1-t_1)}. \qquad (7.5.33)$$

In the special case when the wing is symmetrical about the z-axis, $t_1 = -t_0$ and (7.5.32) can be put into the form

$$w = \frac{U\alpha t_0^2}{BE_1\sqrt{(t_0^2 - x_1^2)}} \quad \text{for } |x_1| < t_0 \text{ on } y_1 = +0, \qquad (7.5.34)$$

where E_1 is the complete elliptic integral of the second kind to modulus $\sqrt{(1-t_0^2)}$. The result (7.5.34) was first obtained by Robinson (1946*b*) and Stewart (1946).

7.6. Superposition of conical fields

The flow over more general plane wings at incidence may be found by suitable superpositions of the conical field solutions found above. These superpositions may be of a finite number of conical fields, or of an infinite number (by integrating solutions with respect to one or more parameters). In the first case all supersonic leading edges and all subsonic edges must be straight. The occurrence of a straight subsonic edge sets a limit to the possible extent of such a wing; such an edge is inside the Mach cone of its upstream end, and a conical field (or a finite combination of conical fields) cannot be found which gives the flow correctly in the influence domain of the portion of this edge which lies inside the Mach cone of an adjacent conical field.

The method of combination may be explained by considering the plane wing shown in fig. 7.10 (over). Fig. 7.10(*a*) shows the mean wing surface in the plane $y=0$. AB, AC are straight supersonic leading edges; BD, CE are straight subsonic edges; AD, AE which meet BD and CE in D and E are the traces on the plane $y=0$ of the Mach cone of A; BG, CF are parts of the traces of the Mach cones of B and C respectively; and DFH, EGH are parts of the traces of the Mach cones of D and E respectively.

As before, since u and w are odd functions, and v is an even function of y, only the solution in $y \geqslant 0$ need be considered. The regions a, b, c, d, e, f, etc., are all regions in $y \geqslant 0$. The region a is the part of the Mach cone of A that is outside the Mach cones of B and C; the region b is bounded by the Mach cones of A and B and the plane characteristic surface through AB, and the region c is defined similarly, with C in place of B. The region d is the part of the Mach cone of B outside the Mach cone of A, and similarly for e, with C in place of B. The region f is inside the Mach cones of A and B, and outside the Mach cones of C and D; g is defined similarly with C, B and E in place of B, C and D, respectively. The region h is inside the Mach cones of B and C, and outside the Mach cones of D and E. Sections through the regions are shown in figs. 7.10(*b*)–7.10(*f*).

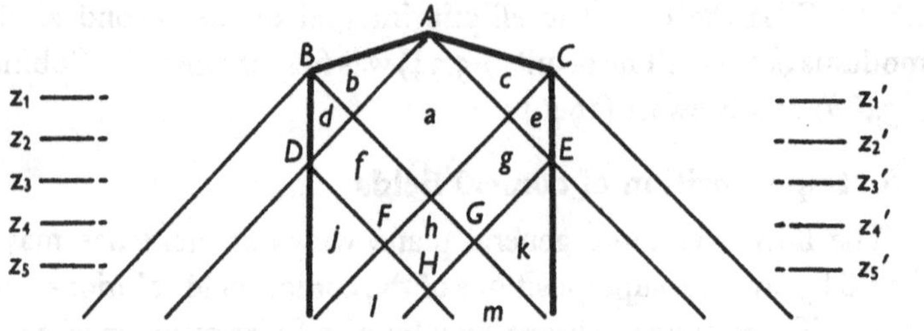

(a) Wing plan form

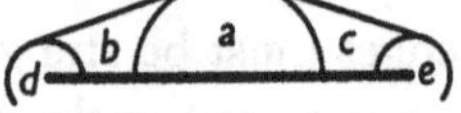

(b) Section on $z_1 z_1'$

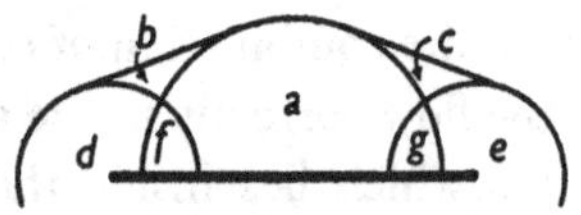

(c) Section on $z_2 z_2'$

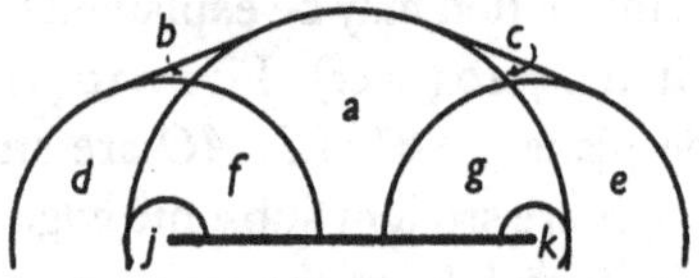

(d) Section on $z_3 z_3'$

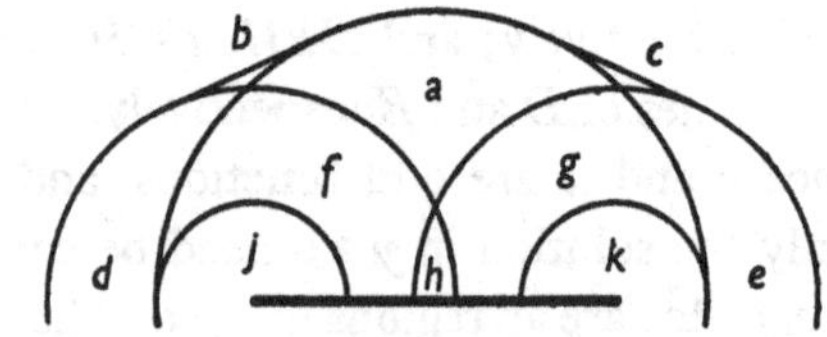

(e) Section on $z_4 z_4'$

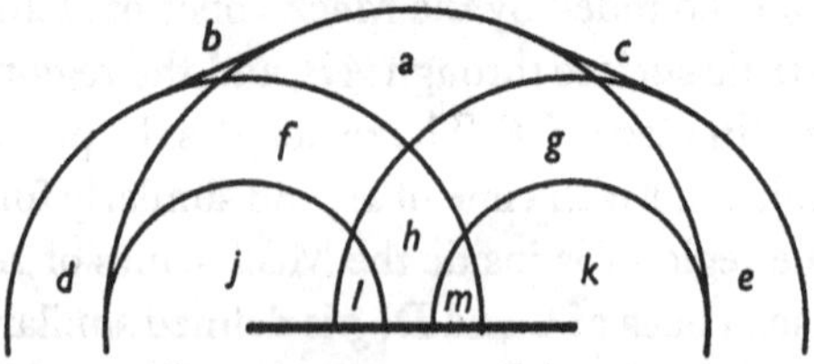

(f) Section on $z_5 z_5'$

Fig. 7.10. (*a*) **Wing planform and characteristic lines in the mean plane of the wing, and** (*b*)–(*f*) **sections of the characteristic surfaces by planes** $z = \text{constant}$.

The velocity fields in the regions a, b, c have been found above, in §§7.2 and 7.5 (i) (since the effects of the breaks in the leading edges at B and C are confined to the interiors of the Mach cones of these points). The velocity fields in the regions d, e are conical fields with origins at B and C respectively, of the type found in §7.5 (iv) (since the effects of the break in the leading edge at A are confined to the interior of the Mach cone of A). The velocity fields in the regions a, b, c, d, e are therefore found as in previous sections of this chapter.

Now denote by $\mathbf{v}_a$, $\mathbf{v}_b$, $\mathbf{v}_c$, $\mathbf{v}_d$, $\mathbf{v}_e$, etc., the solutions for the perturbation velocity in the regions a, b, c, d, e, etc., respectively, with the following stipulations. The velocities $\mathbf{v}_b$ and $\mathbf{v}_c$ are constant, so $\mathbf{v}_b$ and $\mathbf{v}_c$ may be considered to be defined at all points. The solution $\mathbf{v}_a$ may be continued throughout the whole region in which it would apply for an infinite triangular wing, which is the interior of the Mach cone of A, and may be considered to be defined at all points of this region. Similarly, $\mathbf{v}_d$ and $\mathbf{v}_e$ may be continued throughout the interiors of the Mach cones of B and C respectively and may be considered to be defined there. (It is to be understood, however, that $\mathbf{v}_a$, $\mathbf{v}_b$, etc., though now defined in more extended regions, are the solutions for the given wing only in the regions a, b, etc., respectively.) The velocity fields given by $\mathbf{v}_b - \mathbf{v}_d$ and $\mathbf{v}_c - \mathbf{v}_e$ are called 'complementary conical fields' with B and C as origins respectively, and it can easily be shown that the solution in a more extended region inside the Mach cone of A may be obtained by subtracting these complementary fields from the field $\mathbf{v}_a$. In fact the solutions for the regions f, g, h, are respectively

$$\mathbf{v}_f = \mathbf{v}_a - (\mathbf{v}_b - \mathbf{v}_d),$$
$$\mathbf{v}_g = \mathbf{v}_a - (\mathbf{v}_c - \mathbf{v}_e),$$
$$\mathbf{v}_h = \mathbf{v}_a - (\mathbf{v}_b - \mathbf{v}_d) - (\mathbf{v}_c - \mathbf{v}_e).$$

These solutions clearly satisfy the linearized equations. They also satisfy the boundary condition $v = -U\alpha$ on the mean wing surface, and they make the velocity components continuous over the common boundaries of f with a and d, of g with a and e, and of h with f and g. The value of w, and hence of the pressure, on $y = +0$ may now be deduced immediately from the values for the solutions $\mathbf{v}_a$, $\mathbf{v}_b$, $\mathbf{v}_c$, $\mathbf{v}_d$, $\mathbf{v}_e$.

The Mach cones of B and C include portions of the plane $y=0$ that are not on the mean wing surface, and on which the condition $w=0$ must be satisfied to maintain the continuity of pressure. This condition cannot be satisfied by finite combinations of the above type which also satisfy the other conditions. Extended regions of the wing surface can only be treated by superposing continuous distributions of conical fields, or by using the methods of Chapter 6. For precisely the same reason, the solution for the case of § 7.5 (v) cannot be combined with a finite number of conical field solutions to give the flow over an extended form of wing.

The foregoing solution gives the pressure over the whole wing surface if the trailing edge of the wing is any line (not necessarily straight or in straight segments) that lies in the region bounded by *ABDFHGECA* in fig. 7.10 (*a*), and whose angle of sweep nowhere exceeds $\frac{1}{2}\pi-\mu$ (i.e. it must be a supersonic trailing edge).

CHAPTER 8

APPLICATION OF OPERATIONAL METHODS TO SUPERSONIC FLOW

8.1. Heaviside transforms

If $f(z)$ is a function of a real variable, z, such that the integral

$$F(p) = p\int_0^\infty f(z)\,e^{-pz}\,dp \tag{8.1.1}$$

converges when $\Re(p) > A$, where A is some real finite constant, then $F(p)$ is a regular function of the complex variable p in the domain $\Re(p) > A$ which can be continued analytically into the domain $\Re(p) \leqslant A$. The function $F(p)$ is called the Heaviside operational transform, and is of course simply the Laplace transform of $f(z)$ multiplied by p. The advantage of this extra p is that it makes $F(p) = 1$ when $f(z) = 1$, so no symbolic distinction need be made between a function of z and its operational transform in practical application, and (8.1.1) can be written

$$f(z) = F(p) \tag{8.1.2}$$

without fear of confusion.

The relation (8.1.1) can be solved for $f(z)$ to give

$$f(z) = \frac{1}{2\pi i}\int_{\mathscr{L}} F(\lambda)\,e^{\lambda z}\frac{d\lambda}{\lambda}, \tag{8.1.3}$$

where $\mathscr{L}$ is any contour from $A - i\infty$ to $A + i\infty$ which passes to the right of all the singularities of the integrand; this enables $f(z)$ to be determined when $F(p)$ is given.

In the applications made below, it is sometimes found convenient to express $F(p)$ as a product of the form $F_1(p)F_2(p)/p$, and if

$$f_1(z) = F_1(p), \quad f_2(z) = F_2(p), \tag{8.1.4}$$

then it is a simple consequence of (8.1.1) and (8.1.3) that

$$\frac{F_1(p)F_2(p)}{p} = \int_0^z f_1(s)\,f_2(z-s)\,ds, \tag{8.1.5}$$

a result known as the product theorem.

It can also be shown that if $F(p)$ is expanded as an asymptotic series for large values of p, then term-by-term interpretation gives a series expansion of $f(z)$ in ascending powers of z (usually the Taylor series) which converges for sufficiently small values of z, and that if $F(p)$ is expanded as a convergent series for small values of p, then term-by-term interpretation gives an asymptotic series expansion of $f(z)$ for large values of z.

The above brief summary contains all the results required for the applications given below.†

The function $f(z)$ need not be regular, and in fact can be a highly irregular function of the real variable, z, but, as mentioned above, $F(p)$ is always regular when $\Re(p) > A$ and may be continued analytically over the remainder of the p-plane (suitably cut if necessary); thus the transformation method is particularly well adapted for the treatment of linearized supersonic flow problems, in which the potential and the velocity components are piece-wise regular functions of the space variables.‡

If the potential equation for supersonic flow is multiplied by $p\,e^{-pz}$ and integrated between the limits 0 and ∞ with respect to z, it becomes

$$\frac{\partial^2\phi}{\partial x^2}+\frac{\partial^2\phi}{\partial y^2}-B^2p^2\phi=Q(x,y;p)-B^2p^2(\phi)_{z=0}-B^2p\left(\frac{\partial\phi}{\partial z}\right)_{z=0}, \tag{8.1.6}$$

which is the operational form of the equation; here

$$Q(x,y;p)=p\int_0^\infty Q(x,y,z)\,e^{-pz}\,dz. \tag{8.1.7}$$

In all applications made below, $Q=0$ (i.e. there is no source distribution). Also if $z=0$ can be chosen so that the body (or other disturbing influence) lies entirely in $z>0$, then it follows from the boundary condition (ix) of § 4.5, that $\phi=0$ and $\partial\phi/\partial z=0$ on $z=0$. In this case (8.1.6) becomes

$$\frac{\partial^2\phi}{\partial x^2}+\frac{\partial^2\phi}{\partial y^2}-B^2p^2\phi=0. \tag{8.1.8}$$

For cylindrical polar co-ordinates, r, θ, z, defined by

$$x=r\cos\theta,\quad y=r\sin\theta, \tag{8.1.9}$$

† For proofs of these results and a statement of the limitations to which they are subject, the reader is referred to standard text-books on Heaviside or Laplace transforms.

‡ For subsonic flows, where the corresponding quantities are regular functions of the space variables over the whole flow field, Fourier transforms are appropriate if a transformation method is required.

(8.1.6) becomes

$$\frac{\partial^2\phi}{\partial r^2}+\frac{1}{r}\frac{\partial\phi}{\partial r}+\frac{1}{r^2}\frac{\partial^2\phi}{\partial\theta^2}-B^2p^2\phi=0, \qquad (8.1.10)$$

which has the general solution

$$\phi=\sum_{n=0}^{\infty}\{a_n(p)K_n(Bpr)+b_n(p)I_n(Bpr)\}\cos(n\theta+\theta_n), \qquad (8.1.11)$$

where K_n and I_n are modified Bessel functions (Bessel functions of purely imaginary argument) and a_n, b_n and θ_n are functions of p only.

The Bessel functions have the asymptotic expansions†

$$K_n(\lambda)\sim\sqrt{\left(\frac{\pi}{2\lambda}\right)}\,e^{-\lambda}\left\{1+\sum_{m=1}^{\infty}\frac{(n,m)}{(2\lambda)^m}\right\}\quad\text{for } |\arg\lambda|<\tfrac{3}{2}\pi, \qquad (8.1.12)$$

$$I_n(\lambda)\sim\frac{e^{\lambda}}{\sqrt{(2\pi\lambda)}}\left\{1+\sum_{m=1}^{\infty}(-1)^m\frac{(n,m)}{(2\lambda)^m}\right\}$$

$$+\frac{e^{\lambda\pm(n+\frac{1}{2})\pi i}}{\sqrt{(2\pi\lambda)}}\left\{1+\sum_{m=1}^{\infty}\frac{(n,m)}{(2\lambda)^m}\right\}\quad\text{for } |\arg\lambda\mp\tfrac{1}{2}\pi|<\pi \qquad (8.1.13)$$

as $|\lambda|\to\infty$, where

$$(n,m)=\frac{(n+m-\frac{1}{2})!}{m!(n-m-\frac{1}{2})!}. \qquad (8.1.14)$$

It follows from (8.1.12) that the terms in K_n in (8.1.11) represent waves travelling outwards from the z-axis and downstream; the terms in I_n represent waves travelling in both directions, and lead to flows in which the velocity is finite at almost all points on the z-axis.

The pressure is given from the linearized Bernoulli equation, which in operational form is

$$p-p_0=-\rho_0 Up\phi,‡ \qquad (8.1.15)$$

if $\phi=0$ when $z=0$.

8.2. Flow past quasi-cylindrical ducts

A quasi-cylindrical duct is defined to be an axially symmetrical tube of nearly constant radius whose section by a meridian plane has a small slope relative to the tube axis on both external and internal surfaces. If the tube axis is not aligned with the direction of the undisturbed stream, then the inclination must be small if

† G. N. Watson, *The Theory of Bessel Functions* (Cambridge, 2nd ed., 1944).

‡ No confusion should arise from the two uses for p; the pressure nearly always occurs in the combination $p-p_0$.

the flow is to be treated by linearized theory. The tube is assumed to be of finite length with its ends lying in planes normal to the tube axis. The mean surface of the duct can be taken as a cylinder of constant radius, whose axis is parallel to the direction of the undisturbed stream, and whose ends lie in planes normal to its axis. It is further assumed that the internal flow is supersonic for at least a small distance from the upstream end so that no disturbance from the inside can affect the external flow; then the external and internal flows are independent and can be treated separately.

In the following treatment of the flows, the upstream end of the duct is taken to lie in the plane $z=0$ (so there is no disturbance in $z<0$), and the axis of the mean cylindrical surface is taken to be the z-axis. At zero incidence the duct axis coincides with the z-axis, and an incidence α is a rotation through an angle α about the x-axis, in the positive sense, as usual. If $\eta(z)$ is the slope of either the external or internal duct surface relative to the duct axis, and r_0 is the radius of the mean duct surface, then the boundary condition at either the external or internal boundary is

$$(\partial\phi/\partial r)_{r=r_0} = U\{\eta(z) - \alpha\sin\theta\} \quad \text{for } z \geqslant 0. \tag{8.2.1}$$

If $H(p)$ is the operational transform of $\eta(z)$, then this condition can be expressed in operational form as

$$(\partial\phi/\partial r)_{r=r_0} = U\{H(p) - \alpha\sin\theta\}. \tag{8.2.2}$$

For convenience, the radius of the mean surface is taken to be unity ($r_0=1$) and the Mach number of the flow is taken to be $\sqrt{2}$, so that $B=1$. It is an easy matter to transform any given result back to general values of r_0 and B; this is done below for a number of the more important results.

8.3. External flow at zero incidence

In the case of zero incidence, there is no dependence on the polar angle θ, and the external disturbances are propagated outwards and downstream; hence the appropriate potential (with $B=1$) is

$$\phi = a_0(p)\,K_0(pr). \tag{8.3.1}$$

From this and (8.2.2) with $r_0=1$, it follows that

$$[pa_0(p)\,K_0'(pr)]_{r=1} = -pa_0(p)\,K_1(p) = UH(p); \tag{8.3.2}$$

hence
$$\phi = -\frac{K_0(pr)}{pK_1(p)}\, UH(p), \tag{8.3.3}$$
and
$$\frac{\partial\phi}{\partial z} = -\frac{K_0(pr)}{K_1(p)}\, UH(p). \tag{8.3.4}$$

The quantity most usually required is the pressure on the external surface of the duct, and this is given by
$$\left(\frac{p-p_0}{\rho_0 U^2}\right)_{r=1} = \frac{K_0(p)}{K_1(p)}\, H(p). \tag{8.3.5}$$
This operational form can be interpreted by the product theorem, but some care is necessary, since, if $F(\lambda)$ in (8.1.3) is put equal to $\lambda K_0(\lambda)/K_1(\lambda)$, the integral does not converge. The difficulty is overcome by writing (8.3.5) in the form
$$\left(\frac{p-p_0}{\rho_0 U^2}\right)_{r=1} = \left(1 - \frac{K_1(p)-K_0(p)}{K_1(p)}\right) H(p), \tag{8.3.6}$$
and on interpretation by the product theorem, this gives
$$\left(\frac{p-p_0}{\rho_0 U^2}\right)_{r=1} = \eta(z) - \int_0^z W(z-s)\,\eta(s)\,ds, \tag{8.3.7}$$
where
$$W(z) = \frac{p\{K_1(p)-K_0(p)\}}{K_0(p)}. \tag{8.3.8}$$
All the interpreting integrals now converge.

On restoring the general values of r_0 and B, (8.3.7) becomes
$$\left(\frac{p-p_0}{\rho_0 U^2}\right)_{r=r_0} = \frac{1}{B}\left\{\eta(z) - \frac{1}{Br_0}\int_0^z W\left(\frac{z-s}{Br_0}\right)\eta(s)\,ds\right\}. \tag{8.3.9}$$
This result, with $W(z)$ defined by an integral equation, was given by Lighthill (1945); the present derivation by operational methods was given by Ward (1948).

A graph of the function $W(z)$, drawn from tables prepared by the Admiralty Computing Service,† is shown in fig. 8.1. Some of the more interesting properties of $W(z)$ are investigated later, in §8.5.

The pressure for small z (>0) is, from (8.3.9),
$$\left(\frac{p-p_0}{\rho_0 U^2}\right)_{r=r_0} = \frac{\eta(0)}{\sqrt{(M^2-1)}}, \tag{8.3.10}$$
in agreement with the two-dimensional result of §6.1.

† Admiralty Report SRE/ACS 89 (1945) (restricted circulation). These tables are quoted by Ward (1948).

The velocity components at points not on the duct can be found from (8.3.3). In particular, the approximate shape of the streamlines is given from

$$\frac{1}{U}\frac{\partial\phi}{\partial r}=\frac{K_1(pr)}{K_1(p)}H(p), \tag{8.3.11}$$

since $\partial\phi/U\partial r$ gives the approximate slope of the streamlines at any point. By using the asymptotic expansion for K_1, it can be shown that $\partial\phi/\partial r=0$ for $z-B(r-r_0)<0$; and, for $z-B(r-r_0)>0$,

$$\partial\phi/U\partial r=\eta_0(0)\sqrt{(r_0/r)}\{1+O[z-B(r-r_0)]\}. \tag{8.3.12}$$

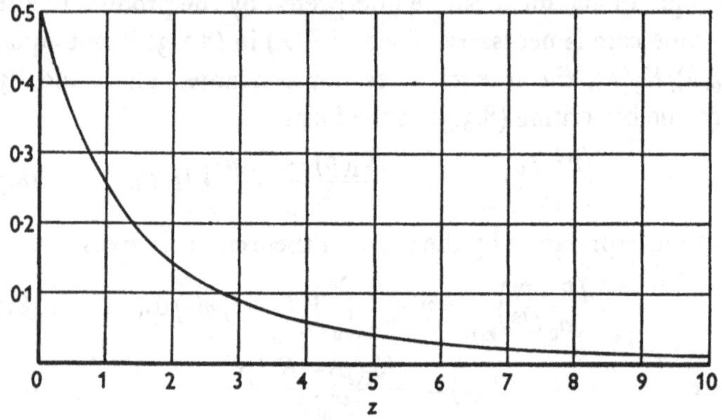

Fig. 8.1. The function $W(z)$, defined by (8.3.8) or (8.5.3).

Thus the deflexion through the initial shock (if $\eta(0)>0$) decreases with increasing r, and it is to be expected that this shock is curved in a more exact representation of the flow.

The external drag coefficient for a quasi-cylindrical duct of length l can be obtained from (8.3.9), and is

$$C_{DO(\text{ext.})}=\frac{\text{external drag}}{\frac{1}{2}\rho_0 U^2\pi r_0^2}=\frac{4}{Br_0}\int_0^l \eta^2(z)\,dz$$
$$-\frac{2}{B^2r_0^2}\int_0^l\int_0^l W\left(\frac{|z-s|}{Br_0}\right)\eta(z)\,\eta(s)\,dz\,ds. \tag{8.3.13}$$

8.4. External flow at incidence

The extra potential due to incidence depends upon θ through the factor $\sin\theta$, as is shown by (8.2.1) or (8.2.2), and the appropriate form for this extra potential is

$$\phi=a_1(p)\,K_1(pr)\sin\theta. \tag{8.4.1}$$

The boundary condition (8.2.2) with $r_0 = 1$ then gives

$$p_1 a_1(p) K_1'(p) = -U\alpha; \tag{8.4.2}$$

hence

$$\phi = -U\alpha \sin\theta \frac{K_1(pr)}{pK_1'(p)}, \tag{8.4.3}$$

and

$$\frac{\partial\phi}{\partial z} = -U\alpha \sin\theta \frac{K_1(pr)}{K_1'(p)}. \tag{8.4.4}$$

The extra pressure on the external surface, due to incidence, is given by

$$\left(\frac{p-p_0}{\rho_0 U^2}\right)_{r=1} = \alpha \sin\theta \frac{K_1(p)}{K_1'(p)}. \tag{8.4.5}$$

For small values of z (>0), the extra pressure is (for general values of r_0 and B)

$$\left(\frac{p-p_0}{\rho_0 U^2}\right)_{r=r_0} = \lim_{\lambda\to\infty} \frac{\alpha \sin\theta}{B} \frac{K_1(\lambda)}{K_1'(\lambda)} = -\frac{\alpha\sin\theta}{\sqrt{(M^2-1)}}, \tag{8.4.6}$$

again in agreement with two-dimensional results.

Equation (8.4.5) may be written in the form

$$\left(\frac{p-p_0}{\rho_0 U^2}\right)_{r=1} = -\alpha \sin\theta \, V_1'(z), \tag{8.4.7}$$

where

$$V_1(z) = -\frac{K_1(p)}{pK_1'(p)}. \tag{8.4.8}$$

For general r_0 and B, (8.4.7) becomes

$$\left(\frac{p-p_0}{\rho_0 U^2}\right)_{r=r_0} = -\alpha \sin\theta \, V_1'\left(\frac{z}{Br_0}\right). \tag{8.4.9}$$

Some properties of $V_1(z)$ and $V_1'(z)$ are investigated in § 8.5, where graphs of the functions are also given.

If a function $M(z)$ is defined by

$$M(z) = \int_0^z V_1'(z)\, z \, dz, \tag{8.4.10}$$

then the lift coefficient, and the moment coefficient about the x-axis, for a duct of length l are respectively

$$C_{L(\text{ext.})} = \frac{\text{external lift}}{\frac{1}{2}\rho_0 U^2 \pi r_0^2} = 2\alpha V_1(l/Br_0), \tag{8.4.11}$$

and

$$C_{M(\text{ext.})} = \frac{\text{external moment}}{\frac{1}{2}\rho_0 U^2 \pi r_0^3} = 2\alpha B M(l/Br_0); \tag{8.4.12}$$

the total external drag coefficient is

$$C_{D(\text{ext.})} = C_{DO(\text{ext.})} + 2\alpha^2 V_1(l/Br_0), \tag{8.4.13}$$

$C_{DO(\text{ext.})}$ being given by (8.3.13).

Some properties of $V_1(z)$ and $M(z)$ are given in §8.5, where it is shown that, for large z,

$$V_1(z) \sim 1 + 1/z^2 + \ldots, \quad M(z) = 2/z + \ldots. \tag{8.4.14}$$

Thus for a long duct ($l > 10Br_0$ say), the external lift and drag coefficients are

$$C_{L(\text{ext.})} \sim 2\alpha,\dagger \quad C_{D(\text{ext.})} \sim C_{DO(\text{ext.})} + 2\alpha^2,\dagger \tag{8.4.15}$$

and the centre of lift is at

$$z \sim 2B^2 r_0^2/l.\dagger \tag{8.4.16}$$

8.5. The functions $W(z)$, $V_n(z)$ and $M(z)$

From (8.3.8), on interpreting by (8.1.3),

$$W(z) = \frac{1}{2\pi i}\int_{\mathscr{L}} \frac{K_1(\lambda) - K_0(\lambda)}{K_1(\lambda)} e^{\lambda z}\, d\lambda. \tag{8.5.1}$$

The Bessel functions $K_0(\lambda)$ and $K_1(\lambda)$ have branch points at $\lambda = 0$, and $K_1(\lambda)$ has no zeros in $|\arg\lambda| < \pi$, hence it follows from the form of the asymptotic expansions (8.1.12) that the contour $\mathscr{L}$ may be transformed into the contour $\mathscr{M}(-\infty, 0+)$. Now let the contour $\mathscr{M}$ become the upper and lower sides of the cut in the λ-plane along the negative real axis, plus a small circle enclosing $\lambda = 0$. Then by integrating along the three parts of the contour, and using the results

$$\left.\begin{aligned} K_0(\lambda e^{\pm i\pi}) &= K_0(\lambda) \mp i\pi I_0(\lambda), \\ K_1(\lambda e^{\pm i\pi}) &= -K_1(\lambda) \pm i\pi I_1(\lambda), \\ I_0(\lambda)K_1(\lambda) + I_1(\lambda)K_0(\lambda) &= 1/\lambda, \end{aligned}\right\} \tag{8.5.2}$$

(8.5.1) gives

$$W(z) = \int_0^\infty \frac{e^{-\lambda z}}{K_1^2(\lambda) + \pi^2 I_1^2(\lambda)} \frac{d\lambda}{\lambda}. \tag{8.5.3}$$

The function $W(z)$ is defined by (8.5.1) only for real $z > 0$, but z may now be taken to be a complex variable, in which case (8.5.3)

† For the corresponding results when the radius of the body is not nearly constant, see §§9.12 and 9.13.

provides an expression for $W(z)$ valid in the domain $\Re(z) > -2$, since the integral converges in this domain.

Analytical continuation of $W(z)$ over the whole z-plane minus a cut along the negative real axis from $z=-2$ to $z=-\infty$ can be effected by rotating the path of integration in (8.5.3).

The nature of the singularity at $z=-2$ can be determined from (8.5.3), and it is found that, as $z \to -2$,

$$W(z) = \frac{2}{\pi}\frac{1}{2+z} + \frac{3}{2\pi}\log\frac{1}{2+z} + O(1), \tag{8.5.4}$$

so $W(z)$ has a simple pole and a logarithmic branch point at $z=-2$.

Substitution of the asymptotic expansion for $K_0(\lambda)$ and $K_1(\lambda)$ in (8.5.1), expansion of the integrand in powers of $1/\lambda$, and integration, gives the Taylor series for small z,

$$W(z) = \tfrac{1}{2} - \tfrac{3}{8}z + \tfrac{3}{16}z^2 - \dots, \tag{8.5.5}$$

which converges for $|z| < 2$.

The asymptotic series for large z can be obtained by substituting the series expansions for the Bessel functions in ascending powers of λ into the contour integral (8.5.1) taken for the contour $\mathscr{M}$, or into the infinite integral (8.5.3), in which the most important contribution is from small λ when z is large. This asymptotic series is

$$W(z) \sim \frac{1}{z^2} + \frac{6}{z^4}(\log 2z - \tfrac{4}{3}) + \dots, \tag{8.5.6}$$

as $|z| \to \infty$, $|\arg z| < \pi$; $\log 2z$ is to be given its principal value. When $z=10$, the terms shown in (8.5.6) give $W(10)=0{\cdot}01099$; the correct value is $0{\cdot}01077$.

Next, consider the functions $V_n(z)$ defined by

$$V_n(z) = -\frac{K_n(p)}{pK_n'(p)} \tag{8.5.7}$$

for z real and >0. The values of the functions and their first derivatives at $z=0$ are given by

$$V_n(0) = \lim_{\lambda\to\infty} -\frac{K_n(\lambda)}{\lambda K_n'(\lambda)} = 0, \tag{8.5.8}$$

and

$$V_n'(0) = \lim_{\lambda\to\infty} -\frac{K_n(\lambda)}{K_n'(\lambda)} = 1. \tag{8.5.9}$$

For the case $n=0$, $K_0'(\lambda)$ has no zeros in $|\arg\lambda|<\pi$, so just as for $W(z)$ above, the contour $\mathscr{L}$ in the interpreting integrals may be replaced by the contour $\mathscr{M}$, and

$$V_0'(z)=\frac{1}{2\pi i}\int_{\mathscr{M}}\frac{K_0(\lambda)}{K_1(\lambda)}e^{\lambda z}\frac{d\lambda}{\lambda}=\int_0^\infty\frac{e^{-\lambda z}}{K_1^2(\lambda)+\pi^2I_1^2(\lambda)}\frac{d\lambda}{\lambda^2}, \quad (8.5.10)$$

and, by integrating,

$$V_0(z)=\int_0^\infty\frac{1-e^{-\lambda z}}{K_1^2(\lambda)+\pi^2I_1^2(\lambda)}\frac{d\lambda}{\lambda^3}, \quad (8.5.11)$$

from which the properties of $V_0(z)$ as a function of a complex variable may be deduced. $V_0(z)$ is related to $W(z)$ by

$$V_0(z)=\int_0^z V_0'(z)\,dz=\int_0^z\left\{1-\int_0^z W(z)\,dz\right\}dz=\int_0^z dz\int_z^\infty W(z)\,dz, \quad (8.5.12)$$

which alternatively enables its properties to be deduced from those of $W(z)$. The Taylor series for small z is

$$V_0(z)=z-\tfrac{1}{4}z^2+\tfrac{1}{16}z^3-\tfrac{1}{64}z^4+\dots, \quad (8.5.13)$$

which converges when $|z|<2$, since $V_0(z)$ has a logarithmic branch point at $z=-2$. The asymptotic expansion for $V_0(z)$ can be obtained by expanding the Bessel functions for small p and interpreting term by term, which gives

$$V_0(z)\sim\log 2z-\frac{1}{z^2}(\log 2z-\tfrac{1}{2})+\dots, \quad (8.5.14)$$

as $|z|\to\infty$ in $|\arg z|<\pi$.

When $n\geqslant 1$, $K_n'(\lambda)$ has zeros in $\frac{1}{2}\pi<|\arg\lambda|<\pi$, and when the contour $\mathscr{L}$ in the interpreting integral for $V_n(z)$ is replaced by the contour $\mathscr{M}$, the residues at the poles of the integrand must be included. For example, $K_1'(\lambda)$ has zeros at $\lambda=-0{\cdot}6453\pm0{\cdot}5012i$, and

$$V_1'(z)=e^{-0{\cdot}6453z}\{1{\cdot}2120\cos 0{\cdot}5012z+0{\cdot}1898\sin 0{\cdot}5012z\} -\frac{1}{2\pi i}\int_{\mathscr{M}}\frac{K_1(\lambda)}{K_1'(\lambda)}e^{\lambda z}\frac{d\lambda}{\lambda}. \quad (8.5.15)$$

By using formulae analogous to (8.5.2), this gives

$$V_1'(z)=e^{-0{\cdot}6453z}\{1{\cdot}2120\cos 0{\cdot}5012z+0{\cdot}1898\sin 0{\cdot}5012z\} -\int_0^\infty\frac{e^{-\lambda z}}{K_1'^2(\lambda)+\pi^2I_1'^2(\lambda)}\frac{d\lambda}{\lambda^2}, \quad (8.5.16)$$

from which the properties of $V_1'(z)$ may be deduced; this result was obtained by the Admiralty Computing Service. A similar expression for $V_1(z)$ can be obtained by integrating (8.5.16) from 0 to z. The Taylor series for small z is

$$V_1(z) = z - \tfrac{1}{4}z^2 - \tfrac{1}{48}z^3 + \dots, \tag{8.5.17}$$

which converges for $|z| < 2$, and the asymptotic expansions for $V_1(z)$ and $V_1'(z)$ as $|z| \to \infty$ in $|\arg z| < \frac{1}{2}\pi$ are

$$V_1(z) \sim 1 + \frac{1}{z^2} + \dots, \quad V_1'(z) \sim -\frac{2}{z^3} + \frac{24}{z^5}(\log 2z - \tfrac{31}{12}) + \dots. \tag{8.5.18}$$

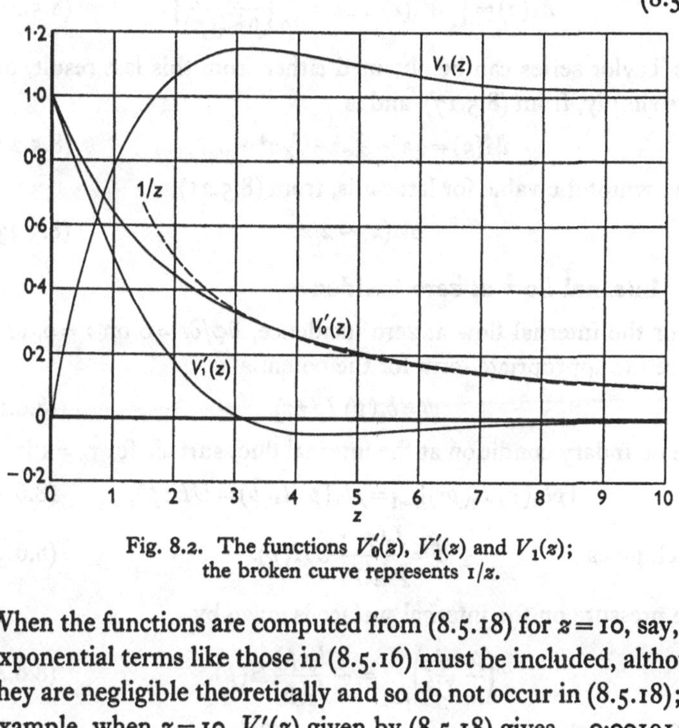

Fig. 8.2. The functions $V_0'(z)$, $V_1'(z)$ and $V_1(z)$; the broken curve represents $1/z$.

When the functions are computed from (8.5.18) for $z = 10$, say, the exponential terms like those in (8.5.16) must be included, although they are negligible theoretically and so do not occur in (8.5.18); for example, when $z = 10$, $V_1'(z)$ given by (8.5.18) gives $-0{\cdot}00191$; by including the exponential terms, the value $-0{\cdot}00163$ is obtained; the correct value of $V_1'(10)$ is $-0{\cdot}00147$.

For other values of n ($\geqslant 1$), it is not difficult to show that

$$V_n(z) \sim \frac{1}{n} + (-1)^{n-1}\frac{(2n-1)!}{2^{2n-2}(n!)^2}\frac{1}{z^{2n}} + \dots, \tag{8.5.19}$$

as $|z| \to \infty$ in $|\arg z| < \frac{1}{2}\pi$.

Curves of $V_0'(z)$, $V_1'(z)$ and $V_1(z)$ are shown in fig. 8.2. Tables of the functions $W(z)$ and $V_1'(z)$ for $z=0(0{\cdot}2)10$ to $5d$ have been computed by the Admiralty Computing Service; these tables are quoted by Ward (1948).

The operational transform of $M(z)$ can be obtained from the result that if $f(z)=F(p)$, then

$$zf(z)=-p\frac{\mathrm{d}}{\mathrm{d}p}\left(\frac{F(p)}{p}\right), \tag{8.5.20}$$

which is easily deduced from (8.1.1). Hence

$$M(z)=\int_0^z V_1'(z)\,z\,\mathrm{d}z=\frac{\mathrm{d}}{\mathrm{d}p}\left(\frac{K_1(p)}{pK_1'(p)}\right). \tag{8.5.21}$$

The Taylor series can be obtained either from this last result, or, more simply, from (8.5.17), and is

$$M(z)=\tfrac{1}{2}z^2-\tfrac{1}{6}z^3-\tfrac{1}{64}z^4+\dots. \tag{8.5.22}$$

The asymptotic value for large z is, from (8.5.21),

$$M(z)\sim 2/z. \tag{8.5.23}$$

8.6. Internal flow at zero incidence

For the internal flow at zero incidence, $\partial\phi/\partial r=0$ on $r=0$, and hence the appropriate form for the potential is

$$\phi=b_0(p)\,I_0(pr). \tag{8.6.1}$$

The boundary condition at the internal duct surface for $r_0=1$ is

$$[pb_0(p)\,I_0'(pr)]_{r=1}=pb_0(p)\,I_1(p)=UH(p), \tag{8.6.2}$$

which gives

$$\phi=\frac{I_0(pr)}{pI_1(p)}\,UH(p). \tag{8.6.3}$$

The pressure on the internal surface is given by

$$\left(\frac{p-p_0}{\rho_0 U^2}\right)_{r=1}=-\frac{I_0(p)}{I_1(p)}H(p). \tag{8.6.4}$$

If the product theorem is applied to (8.6.4), in a similar way to that used in § 8.3, the formula (8.1.3) gives a divergent integral for the interpretation of $p\{I_0(p)-I_1(p)\}/I_1(p)$. This is due to the more complicated behaviour of $I_0(p)$ and $I_1(p)$ at infinity (cf. the asymptotic series for $K_n(\lambda)$ and $I_n(\lambda)$ in (8.1.12) and (8.1.13)). The difficulty can be overcome by defining a new function $\eta_1(z)$ such that

$$\eta(z)=\eta(0)+\eta_1(z),\quad \eta_1(z)=H_1(p); \tag{8.6.5}$$

(8.6.4) can then be written

$$\left(\frac{p-p_0}{\rho_0 U^2}\right)_{r=1} = -\left\{H(p) - \frac{I_1(p)-I_0(p)}{I_1(p)}\,\eta(0) - \frac{I_1(p)-I_0(p)}{pI_1(p)}\,pH_1(p)\right\}. \tag{8.6.6}$$

All the operational transforms now have a meaning, and (8.6.6) gives

$$\left(\frac{p-p_0}{\rho_0 U^2}\right)_{r=1} = -\left\{\eta(z) - \eta(0)\,T(z) - \int_0^z \eta_1'(z)\,T(z-s)\,ds\right\}, \tag{8.6.7}$$

where

$$T(z) = \frac{I_1(p)-I_0(p)}{I_1(p)}. \tag{8.6.8}$$

From (8.6.8), $T(0)=0$, and hence, by integrating (8.6.7) by parts,

$$\begin{aligned}\left(\frac{p-p_0}{\rho_0 U^2}\right)_{r=1} &= -\left\{\eta(z) - \eta(0)\,T(z) - \int_0^z \eta_1(s)\,T'(z-s)\,ds\right\}\\ &= -\left\{\eta(z) - \int_0^z \eta(s)\,T'(z-s)\,ds\right\}.\end{aligned} \tag{8.6.9}$$

The interpreting integral for $T(z)$ is

$$T(z) = \frac{1}{2\pi i}\int_{\mathscr{L}} \frac{I_1(\lambda)-I_0(\lambda)}{I_1(\lambda)}\,e^{\lambda z}\,\frac{d\lambda}{\lambda}. \tag{8.6.10}$$

If the asymptotic series for $I_0(\lambda)$ and $I_1(\lambda)$ are now substituted in the integrand, then it can be expanded formally in powers of $e^{-\lambda}$ and interpreted term by term. Each term in the integrand has a factor $e^{\lambda(z-2n)}$ say, and the corresponding term in the interpretation vanishes for $z<2n$. Thus there is always a finite number of terms, and the question of convergence does not arise. In particular, for $z<2$, only the first series in (8.1.13) occurs, and by comparing this with (8.1.12), it will be seen that the series for $T'(z)$ near the origin is the same as that for $-W(-z)$. Hence

$$T'(z) = -W(-z) \quad \text{for } 0<z<2.\dagger \tag{8.6.11}$$

From (8.6.9) and (8.6.11), it follows that, for general values of r_0 and B,

$$\left(\frac{p-p_0}{\rho_0 U^2}\right)_{r=r_0} = -\frac{1}{B}\left\{\eta(z) + \frac{1}{Br_0}\int_0^z W\left(\frac{s-z}{Br_0}\right)\eta(s)\,ds\right\}, \tag{8.6.12}$$

† This result has been proved rigorously by Professor S. Goldstein (private communication).

and there results an expression for the internal drag coefficient at zero incidence similar to that given in (8.3.13) for the external drag,

$$C_{DO(\text{int.})} = \frac{4}{Br_0}\int_0^l \eta^2(z)\,dz + \frac{2}{B^2r_0^2}\int_0^l\int_0^l W\left(-\frac{|z-s|}{Br_0}\right)\eta(z)\,\eta(s)\,dz\,ds, \tag{8.6.13}$$

valid for $0 < l < 2Br_0$.

If $T(z)$ is required for $z > 2$, then (8.6.10) must be expanded by the residue theorem, and the resulting series computed. $T(z)$ is defined only for real $z \geqslant 0$, and its behaviour is rather complicated, since it has certain periodically distributed singularities and discontinuities whose forms are discussed in the next section. That the results (8.6.12) and (8.6.13) are valid only for $l < 2Br_0$ is not a very serious practical disadvantage, since the flow does not usually remain supersonic for very far down the inside of the duct, but becomes subsonic through a system of shocks. Moreover, in general the pressure on the internal surface given by (8.6.12) becomes logarithmically infinite at $z = 2Br_0$, and the approximations of linearized theory are no longer valid.

8.7. The singularities and discontinuities in $T(z)$

The contour integral (8.6.10) for $T(z)$ may be evaluated by the residue theorem. For $z < 0$, the integral vanishes, while for $z > 0$, since the only poles of the integrand are at $\lambda = \pm i j_{1,n}$ $(n = 0, 1, 2, \ldots)$, where $j_{1,n}$ is the nth positive zero of $J_1(\lambda)$ $(j_{1,0} = 0)$,

$$T(z) = 1 - 2z - 2\sum_{n=1}^{\infty} \frac{J_0(j_{1,n})}{j_{1,n}J_1'(j_{1,n})}\sin j_{1,n}z. \tag{8.7.1}$$

It does not seem to be possible to sum this series as a closed expression, but it is possible to extract the singular and discontinuous parts, the remainder being a continuous bounded function throughout the field.

The roots $j_{1,n}$ have asymptotic expansions as functions of n, given by

$$j_{1,n} \sim (n+\tfrac{1}{4})\pi - \frac{3}{8(n+\frac{1}{4})\pi} + \ldots, \tag{8.7.2}$$

and the Bessel functions $J_0(z), J_1'(z)$ have the asymptotic expansions

$$J_0(\lambda) \sim \sqrt{(2/\pi\lambda)}\left\{\cos(\lambda - \tfrac{1}{4}\pi) + \frac{1}{8\lambda}\sin(\lambda - \tfrac{1}{4}\pi) + \ldots\right\},$$

$$J_1'(\lambda) \sim \sqrt{(2/\pi\lambda)}\left\{\cos(\lambda - \tfrac{1}{4}\pi) - \frac{7}{8\lambda}\sin(\lambda - \tfrac{1}{4}\pi) + \ldots\right\}. \tag{8.7.3}$$

Therefore, for large n,

$$\frac{J_0(j_{1,n})}{j_{1,n}J_1'(j_{1,n})}\sin j_{1,n}z \sim \frac{1}{n\pi}\sin(n+\tfrac{1}{4})\pi z + O(1/n^2) \tag{8.7.4}$$

uniformly in z, and the series

$$\sum_{n=1}^{\infty}\left\{\frac{J_0(j_{1,n})}{j_{1,n}J_1'(j_{1,n})}\sin j_{1,n}z - \frac{1}{n\pi}\sin(n+\tfrac{1}{4})\pi z\right\} \tag{8.7.5}$$

converges uniformly and absolutely. Hence, any singularities and discontinuities in $T(z)$ must be contained in the series

$$-\frac{2}{\pi}\sum_{n=1}^{\infty}\frac{1}{n}\sin(n+\tfrac{1}{4})\pi z = -\frac{2}{\pi}\sin\frac{\pi z}{4}\sum_{n=1}^{\infty}\frac{1}{n}\cos n\pi z$$
$$-\frac{2}{\pi}\cos\frac{\pi z}{4}\sum_{n=1}^{\infty}\frac{1}{n}\sin n\pi z = -\frac{2}{\pi}\sin\frac{\pi z}{4}C(z) - \frac{2}{\pi}\cos\frac{\pi z}{4}S(z), \text{ say.} \tag{8.7.6}$$

Now $\quad C(z)+\mathrm{i}S(z) = \sum_{n=1}^{\infty} e^{\mathrm{i}n\pi z}/n = -\log(1-e^{\mathrm{i}\pi z})$

$$= -\tfrac{1}{2}\log\{2(1-\cos\pi z)\} + \mathrm{i}\tan^{-1}\left(\frac{\sin\pi z}{1-\cos\pi z}\right), \tag{8.7.7}$$

so $C(z)$ and $S(z)$ have logarithmic singularities and discontinuities respectively at $z=0, \pm 2, \pm 4, \ldots$.

It follows from these results that $T(z)$ is continuous at $z=0$ ($T(0)=0$, and $T(z)=0$ for $z<0$); at $z=2$, $T(z)$ has a singularity $(2/\pi)\log|z-2|$; at $z=4$, $T(z)$ has a discontinuity,

$$T(4+0)-T(4-0)=2;$$

at $z=6$, $T(z)$ has a singularity $-(2/\pi)\log|z-4|$, etc.

8.8. The singularities in the internal flow at zero incidence

If $\eta(z)$, the slope of the internal surface, has a discontinuity, or a non-zero value at $z=0$, singularities and discontinuities occur in the internal flow. It will be seen that the same expression, (8.6.3), for the potential will arise if the flow is calculated for a circular pipe whose cross-section is uniform for $z<0$, and which contains a uniform supersonic flow in this region. The existence of singularities in such a flow was noticed by Ward (1945). R. E. Meyer (1948), in his investigations of the exact solutions with axial symmetry, found similar singularities, and called the general phenomenon the 'radial focusing effect', because disturbances which originate, say, on the wall in pipe flow increase in magnitude on the downstream characteristic cone through the disturbance as the axis is approached. Meyer considered discontinuities of curvature rather than discontinuities of slope, but the difference is not important mathematically in linearized theory.

In order to simplify the investigation, it is convenient to take $\eta(z)=0$ for $z\leqslant 0$, and $\eta(z)=\eta_0=$ *constant* for $z>0$; the expressions for the velocity components then become (for $B=1$)

$$\frac{\partial\phi}{\partial r}=U\eta_0\frac{I_1(pr)}{I_1(p)},\quad \frac{\partial\phi}{\partial z}=U\eta_0\frac{I_0(pr)}{I_1(p)}. \tag{8.8.1}$$

When $z+r-1$ is small and greater than zero, the expressions in (8.8.1) may be interpreted by replacing the Bessel functions by their asymptotic expansions to give

$$\begin{aligned} &\partial\phi/\partial r=U\eta_0/\sqrt{r}, && \partial\phi/\partial z=U\eta_0/\sqrt{r} && \text{for } z+r-1>0,\ r\neq 0,\\ &\partial\phi/\partial r=0, && \partial\phi/\partial z=U\eta_0/\sqrt{(z-1)} && \text{for } z-1>0,\ r=0. \end{aligned} \tag{8.8.2}$$

Thus on the axis just downstream from the characteristic cone from the discontinuity (Meyer's 'leading characteristic') there are singularities in the velocity components, and, in addition, there are discontinuities on the characteristic cone itself which increase in magnitude as the axis is approached. This is the linearized theory of radial focusing.

The singularities downstream from the leading characteristic can be found by interpreting the operational transforms in (8.8.1) by using the residue theorem on their integral interpretations. When $z+r-1<0$, both integrals vanish, while for $z+r-1>0$,

$$\frac{\partial\phi}{\partial r}=2U\eta_0\left\{\tfrac{1}{2}r+\sum_{n=1}^{\infty}\frac{J_1(j_{1,n}r)}{j_{1,n}J_1'(j_{1,n})}\cos j_{1,n}z\right\}, \tag{8.8.3}$$

$$\frac{\partial\phi}{\partial z}=2U\eta_0\left\{z+\sum_{n=1}^{\infty}\frac{J_0(j_{1,n}r)}{j_{1\,n}J_1'(j_{1\,n})}\sin j_{1,n}z\right\}, \tag{8.8.4}$$

where $j_{1,n}$ is the nth root of $J_1(\lambda)$, as in § 8.7. Of course neither of these series expressions is valid in $z+r-1<0$, although both continue into this region.

The singularities and discontinuities in these series can be extracted by the method used in § 8.7 for $T(z)$.† The most noteworthy result of the investigation is that the discontinuities are of

† For full details and results see Ward (1948).

opposite natures on $z+r=1$ and $z-r=3$, $z+r=5$; thus if $z+r=1$ is an expansion, then $z-r=3$, $z+r=5$ are compressions, and vice versa. This is in direct contrast with the results of linearized theory for two-dimensional flow of a similar nature, in which there are no singularities, and all the discontinuities (on $z \pm r = 1, 3, 5$, etc.) are of the same kind, that is, all expansive or all compressive. The fact that the compressive discontinuities become stronger as the axis is approached suggests that a flow of this kind should contain normal disc shocks (Mach shocks) on the axis; such shocks could certainly not be represented by solutions of the linearized equations.

The results given above apply to the case when there is a discontinuity of wall slope at $z=0$, but similar results can be found for other types of discontinuity in the wall. In particular for discontinuous curvature at $z=0$ (of amount η_0), the results for $\partial^2\phi/\partial z^2$ and $\partial^2\phi/\partial r\,\partial z$ are the same as those for $\partial\phi/\partial z$ and $\partial\phi/\partial r$ respectively, and the results for $\partial^2\phi/\partial r^2$ are similar. The results obtained by Meyer on the leading characteristic agree with these if the derivatives are taken with respect to distance between successive characteristics instead of co-ordinate distance. In more exact theory, of course, the characteristics are not parallel as they are in linearized theory. Although the velocity components are finite everywhere for discontinuous curvature, the validity of the linearized solution as a first approximation to the exact solution on and near the characteristics $z \pm r = 1, 3, 5$, etc., is doubtful, since Meyer has shown that the series of successive approximations obtained by substituting the linearized solution in the exact potential equation (for a perfect gas with constant specific heats), and working out the correction terms, does not converge on the first reflected characteristic, $z-r=1$. That is to say, the series expansions for the exact potential in powers of the value of the discontinuity in curvature (analogous to the expansion in powers of t given in (1.8.4)) does not converge uniformly at $z-r=1$ (nor on the other characteristics where there are singularities). Thus it is not possible to say whether or not the linearized solution is a valid continuation through $z-r=1$ and subsequent singularities. The problem of giving a full discussion of flow, with singularities, inside a round tube is one of very considerable mathematical difficulty, and has not yet been solved. When the solution is found, it will be interesting to make a comparison with the linearized solution, and to determine whether or not the linearized solution is a valid first approximation outside neighbourhoods of the singularities.

8.9. Internal flow at incidence

The results for the internal flow at incidence are very similar to those obtained already, and are not of sufficient importance to merit a detailed discussion. For a duct of radius r_0, it can be shown that the extra pressure on the internal surface due to incidence α is given by

$$\left(\frac{p-p_0}{\rho_0 U^2}\right)_{r=r_0} = \frac{\alpha \sin\theta}{B} V_1'\left(-\frac{z}{Br_0}\right) \quad \text{for } 0<z<2Br_0, \tag{8.9.1}$$

from which the internal lift coefficient, the moment coefficient about the x-axis, and the drag coefficient can be obtained for a duct of length less than $2Br_0$.

8.10. Axially symmetrical supersonic free jets

If a uniform supersonic stream is flowing in a circular pipe of uniform cross-section which terminates at $z=0$, so that the stream emerges into a fluid at rest with uniform pressure p_1, then an axially symmetrical free jet is formed. On the assumption of inviscid flow, the boundary of the jet is a vortex sheet, and the boundary condition is that the pressure should be continuous. If the pressure in the pipe (in $z<0$) is p_0, then the flow in the jet can be treated by linearized theory when $(p_1-p_0)/\rho_0 U^2$ is small compared with unity.

As before, the Mach number is taken to be $\sqrt{2}$, and the radius of the pipe is taken to be unity. The appropriate form of the potential is

$$\phi = b_0(p)\, I_0(pr), \tag{8.10.1}$$

and the boundary condition on the mean jet boundary, $r=1$, is

$$\left(\frac{\partial\phi}{\partial z}\right)_{r=1} = \frac{p_0-p_1}{\rho_0 U} \tag{8.10.2}$$

from condition (ii) of § 4.5. Thus the potential is

$$\phi = \frac{p_0-p_1}{\rho_0 U}\frac{I_0(pr)}{pI_0(p)}. \tag{8.10.3}$$

The most interesting property of such a jet is the shape of the boundary, which can be determined from $(\partial\phi/U\partial r)_{r=1}$, since this quantity gives the approximate slope of the bounding streamline. From (8.10.3),

$$\frac{1}{U}\left(\frac{\partial\phi}{\partial r}\right)_{r=1}=\frac{p_0-p_1}{\rho_0 U^2}\frac{I_1(p)}{I_0(p)}, \tag{8.10.4}$$

and, since $r=1$ when $z=0$, the boundary profile is given by

$$\begin{aligned} r &= 1+\frac{1}{U}\int_0^z\left(\frac{\partial\phi}{\partial r}\right)_{r=1}dz, \quad z\geqslant 0, \\ &= 1+\frac{p_0-p_1}{\rho_0 U^2}F(z), \end{aligned} \tag{8.10.5}$$

where
$$F(z)=\frac{1}{2\pi i}\int_{\mathscr{L}}\frac{I_1(\lambda)}{I_0(\lambda)}e^{\lambda z}\frac{d\lambda}{\lambda}. \tag{8.10.6}$$

The contour integral in (8.10.6) can be evaluated by the residue theorem. If $j_{0,n}$ ($n=1, 2, 3$, etc.) is the nth positive zero of $J_0(\lambda)$, then the poles of the integrand are at $\lambda=0$, $\lambda=\pm ij_{0,n}$, and, for $z>0$,

$$F(z)=2\sum_{n=1}^{\infty}(1/j_{0,n}^2)\cos j_{0,n}z-\tfrac{1}{2}. \tag{8.10.7}$$

A graph of $F(z)$ prepared by the Admiralty Computing Service is shown in fig. 8.3. It will be seen that $F(z)$ is quasi-periodic, has periodically distributed discontinuities of slope, and that for $z>0$, $0<F(z)<1$, so the jet never returns to its original diameter.

If the jet has a mean radius r_0, and Mach number M, then the jet profile is given by

$$r=r_0+\frac{p_0-p_1}{\rho_0 U^2}B^2 r_0 F\left(\frac{z}{Br_0}\right). \tag{8.10.8}$$

The flow inside the jet can be investigated, and it is found that there are periodically distributed discontinuities and singularities in the velocity components, quite analogous to those found for pipe flow above. These account for the rather surprising shape of the jet profile, and cast considerable doubt on the validity of the solution. The jet profile in the corresponding two-dimensional problem is shown in fig. 8.4 for comparison with that for the axially symmetrical case. Only those portions of the profiles which are close to the jet orifice can be observed in practice, since mixing occurs to destroy the sharpness of the vortex layer, but such experiments as can be carried out seem to give at least qualitative agreement with the two pictures. The dotted lines in fig. 8.4 represent characteristics, and both cases are for $p_0 > p_1$.

A simpler approximation to the solution of this problem was given by Prandtl (1904); the results given above were obtained by Taunt and Ward (1946*a*), using an integral equation method.

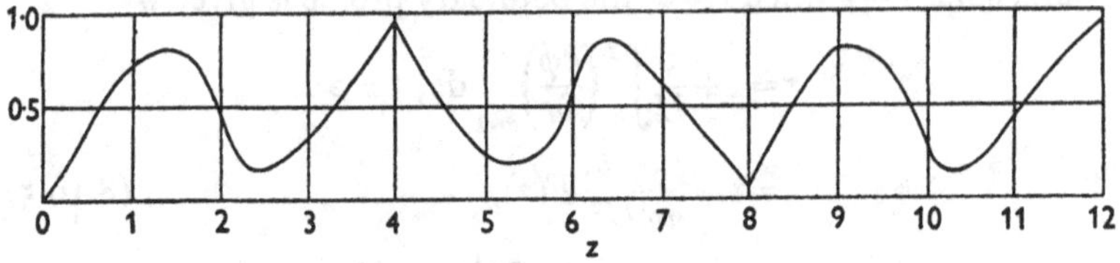

Fig. 8.3. The function $F(z)$ defined by (8.10.7).

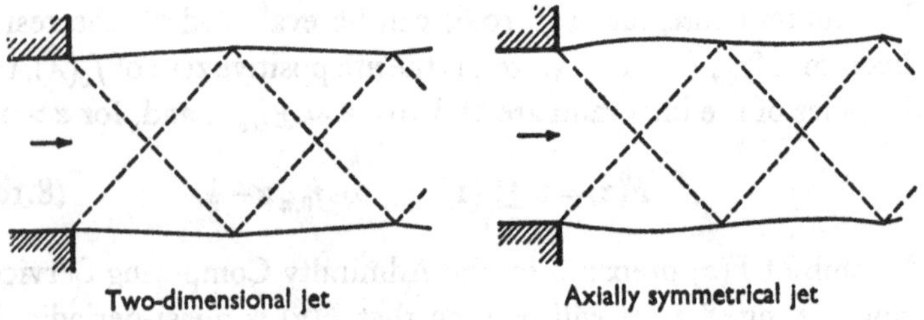

Fig. 8.4. A comparison of the profiles of two-dimensional and axially symmetrical supersonic free jets emerging into lower pressure regions.

8.11. Plane wings

Operational methods have been applied to the solution of the problem of lifting plane wings at supersonic speeds by Gunn (1947) in a paper of very great mathematical interest, but the greater simplicity of the methods of Chapters 6 and 7 has caused these latter to be preferred for practical applications.† However, the operational approach has one advantage over the direct application of Evvard's method, inasmuch as it gives results for wings of small aspect ratio more easily; this has been demonstrated by Stewartson (1950), who has given the asymptotic value of the pressure on a plane wing of rectangular planform for large values of z/B. It should be possible to solve this problem completely by transforming to

† It is interesting to note in passing that Gunn obtained formulae for plane wings that are equivalent to Evvard's, although expressed in very different form.

elliptic cylindrical co-ordinates, but the theory of Mathieu functions is not sufficiently developed to make this approach profitable, although the asymptotic solution can be derived in this way. Stewartson obtained the appropriate solution of (8.1.8) by superposing solutions of the form $K_1(Bpr)\sin\theta$; in terms of rectangular cartesian co-ordinates, this gives

$$\phi=\int_{-b_1(p)}^{b_2(p)}\frac{BpyF(x')}{\sqrt{\{(x-x')^2+y^2\}}}K_1(Bp\sqrt{\{(x-x')^2+y^2\}})\,dx' \tag{8.11.1}$$

as a possible solution of (8.1.8), where $F(x')$ is a function of x' and p, and $b_1(p)$, $b_2(p)$ are functions of p only.

For a plane wing of semi-span b at incidence α, whose mean surface in $y=0$ is $|x|<b$, $z>0$, the boundary conditions are (i) $\phi=0$ and $\partial\phi/\partial z=0$ for $z\leqslant 0$, which is already implicit in the above solution, (ii) $\phi=0$ on $y=0$ for $|x|>b$, $z>0$, and (iii) $\partial\phi/\partial y=-U\alpha$ on the mean wing surface. If in (8.11.1) b_1 and b_2 are put equal to b, and $F(x)=(U\alpha/\pi)f(x)$, then

$$\phi=\frac{U\alpha}{\pi}\int_{-b}^{b}\frac{Bpyf(x')}{\sqrt{\{(x-x')^2+y^2\}}}K_1(Bp\sqrt{\{(x-x')^2+y^2\}})\,dx', \tag{8.11.2}$$

and
$$\lim_{y\to 0}\phi=\lim_{y\to 0}\frac{U\alpha}{\pi}\int_{-b}^{b}\frac{yf(x')}{(x-x')^2+y^2}\,dx'$$

$$=\left.\begin{matrix}U\alpha\,\mathrm{sgn}\,(y)f(x), & \text{for } |x|<b\\ 0, & \text{for } |x|>b\end{matrix}\right\}, \tag{8.11.3}$$

which shows that (8.11.2) satisfies the boundary condition (ii) above; also since ϕ is continuous everywhere, $f(\pm b)=0$. The boundary condition (iii) is satisfied if

$$\left(\frac{\partial\phi}{\partial y}\right)_{y=0}=\frac{U\alpha}{\pi}\int_{-b}^{b}\frac{f'(x')}{x'-x}\,dx'$$
$$+\frac{U\alpha}{\pi}\int_{-b}^{b}f(x')\left(\frac{BpK_1(Bp|x-x'|)}{|x-x'|}-\frac{1}{(x-x')^2}\right)dx'=-U\alpha, \tag{8.11.4}$$

which is an integral equation for $f(x)$. This can be solved for $f(x)$ to give results equivalent to those obtained by direct application of the methods of Chapters 6 and 7, although the analysis is rather lengthy. The asymptotic form of solution for large values of Bz can be obtained by considering the expansion of (8.11.4) for small values of p; when $p=0$ the integrand in the second integral vanishes, and (8.11.4) becomes

$$\frac{1}{\pi}\int_{-b}^{b}\frac{f'(x')}{x-x'}\,dx'=1, \tag{8.11.5}$$

which has the solution

$$f(x)=\sqrt{(b^2-x^2)}. \tag{8.11.6}$$

The asymptotic values of the lift, drag and moment coefficients for a wing of very small aspect-ratio can now be found easily; for, as *chord/B* becomes very large, the lift coefficient, C_L, is given by

$$C_L=\frac{\text{lift}}{\frac{1}{2}\rho_0 U^2(\text{wing area})}\sim\frac{4/U}{\text{wing area}}\lim_{p\to 0}\int_{-b}^{b}(\phi)_{y=+0}\,dx=\tfrac{1}{2}\pi A\alpha, \tag{8.11.7}$$

where A is the aspect-ratio; the drag coefficient $C_D\sim\frac{1}{2}\pi A\alpha^2$, and the moment about the leading edge vanishes.

Stewartson obtained better approximations for $f(x)$ by assuming an expansion of the form

$$f(x)=\sum_{n=0}^{\infty}a_n(1-x^2/b^2)^{n+\frac{1}{2}}, \tag{8.11.8}$$

where the coefficients a_n are functions of p, and he determined a_0, a_1, a_2 and a_3 as quotients of power series in p and $\log p$, the denominators being the same in all coefficients. The interpretation of (8.11.8) follows the lines of the analysis given in § 8.5, the principal difficulty being the determination of the poles arising from the zeros of the common denominator. Stewartson computed four of these zeros, and found that good agreement with the known exact linearized solution for values of z greater than about $1\cdot 5bB$ could be obtained by taking account only of those two zeros nearest to the origin. The inclusion of the other two made the agreement worse. He conjectured that there is an infinite number of such zeros, which is confirmed by noticing that the common denominator must be a factor in a Mathieu function.

PART III SLENDER-BODY THEORY

CHAPTER 9

FLOW AT AND NEAR THE SURFACES OF SLENDER BODIES

9.1. The linearized potential for axially symmetrical subsonic flow past a body of revolution

The incompressible flow past a body of revolution can be represented by a suitable distribution of sources inside the body, and, if the body is slender and has a smooth meridian profile, it is usually assumed that a source distribution on the axis suffices for the representation of the flow outside the body. It follows from the transformation theory of § 2.2 that it should be possible to represent the linearized subsonic compressible flow by an axial distribution of subsonic sources.

Consider a body of revolution which is pointed at both ends and take the z-axis to coincide with the body axis, the origin of co-ordinates to be located at the upstream pointed end, and the unit of length to be so adjusted that the downstream pointed end is at $z=1$. Then, for an axial distribution of sources, the perturbation potential has the form

$$\phi = \int_0^1 \frac{f(s)}{\sqrt{\{(z-s)^2+\beta^2 r^2\}}}\,ds, \tag{9.1.1}$$

where r is the radius in cylindrical polar co-ordinates ($r^2=x^2+y^2$).

Let the body surface be given by

$$r=R(x) \quad \text{for } 0 \leqslant z \leqslant 1, \tag{9.1.2}$$

where $R(z)$ is a piece-wise analytic function of z which satisfies the conditions

$$R(z) < Atz(1-z), \quad |R'(z)| < At, \quad |R''(z)| < At, \quad |R'''(z)| < At, \tag{9.1.3}$$

t being the thickness ratio of the body, defined as $\max\{2R(z)\}$, and A being some constant which is independent of t; for linearized theory to be applicable, t must be small compared with unity. The boundary condition at the body surface is then

$$(\partial\phi/\partial r)_{r=R(z)} = UR'(z) \quad \text{for } 0 \leqslant z \leqslant 1, \tag{9.1.4}$$

from which the unknown function $f(s)$ in (9.1.1) is to be determined.

In order to differentiate (9.1.1) with respect to r, it is convenient to integrate by parts first; then

$$\phi = f(1)\sinh^{-1}\left(\frac{1-z}{\beta r}\right) + f(0)\sinh^{-1}\left(\frac{z}{\beta r}\right) - \int_0^1 f'(s)\sinh^{-1}\left(\frac{s-z}{\beta r}\right) ds, \tag{9.1.5}$$

and hence

$$\frac{\partial\phi}{\partial r} = -f(1)\frac{1-z}{r\sqrt{\{(1-z)^2+\beta^2r^2\}}} - f(0)\frac{z}{r\sqrt{\{z^2+\beta^2r^2\}}} + \int_0^1 \frac{(s-z)f'(s)}{r\sqrt{\{(z-s)^2+\beta^2r^2\}}} ds. \tag{9.1.6}$$

When $r=R(z)$, it is not difficult to show that

$$(r\,\partial\phi/\partial r)_{r=R(z)} = -2f(z) + \max\{|f(1)|, |f(0)|\}\,O(t^2) + \max|f'(z)|\,O(t^2\log t), \tag{9.1.7}$$

and the boundary condition (9.1.4) gives

$$(r\,\partial\phi/\partial r)_{r=R(z)} = UR(z)R'(z) = (U/2\pi)S'(z), \tag{9.1.8}$$

where $S(z)$ $(=\pi\{R(z)\}^2)$ is the cross-sectional area of the body. Therefore

$$f(z) = -(U/4\pi)S'(z)\{1+O(t^2\log t)\}, \tag{9.1.9}$$

since $S'(0)=S'(1)=0$ and $S''(z)=O(t^2)$. Factors $1+O(t)$ are already neglected in the derivation of the linearized equations, so it appears from (9.1.9) that the potential can be written

$$\phi = -\frac{U}{4\pi}\int_0^1 \frac{S'(s)}{\sqrt{\{(z-s)^2+\beta^2r^2\}}} ds, \tag{9.1.10}$$

without making further errors in approximation. On integrating (9.1.10) by parts,

$$\phi = -\frac{U}{4\pi}\int_0^1 S''(s)\sinh^{-1}\left(\frac{z-s}{\beta r}\right) ds, \tag{9.1.11}$$

and by differentiating with respect to z and r, this gives

$$\frac{\partial\phi}{\partial z} = -\frac{U}{4\pi}\int_0^1 \frac{S''(s)}{\sqrt{\{(z-s)^2+\beta^2r^2\}}} ds, \tag{9.1.12}$$

and

$$\frac{\partial\phi}{\partial r} = -\frac{U}{4\pi r}\int_0^1 \frac{(s-z)S''(s)}{\sqrt{\{(z-s)^2+\beta^2r^2\}}} ds. \tag{9.1.13}$$

For $0 \leqslant z \leqslant 1$, (9.1.11) can be written

$$\phi = \frac{U}{4\pi}\left\{\int_z^1 S''(s)\sinh^{-1}\left(\frac{s-z}{\beta r}\right) ds - \int_0^z S''(s)\sinh^{-1}\left(\frac{z-s}{\beta r}\right) ds\right\}, \tag{9.1.14}$$

and since, when r is small compared with $|s-z|$,

$$\sinh^{-1}\left(\frac{|s-z|}{\beta r}\right)-\log\left(\frac{2|s-z|}{\beta r}\right)=O\left(\frac{r^2}{(s-z)^2}\right), \quad (9.1.15)$$

an approximation to ϕ near the body (for small r) can be obtained by replacing the inverse hyperbolic sines by logarithms. Now

$$\left|\phi-\frac{U}{4\pi}\left\{\int_z^1 S''(s)\log\left(\frac{2(s-z)}{\beta r}\right)ds-\int_0^z S''(s)\log\left(\frac{2(z-s)}{\beta r}\right)ds\right\}\right|$$
$$\leqslant UA^2t^2\int_0^1\left|\sinh^{-1}\left(\frac{|s-z|}{\beta r}\right)-\log\left(\frac{2|s-z|}{\beta r}\right)\right|ds, \quad (9.1.16)$$

and the integral from $s=0$ to $s=1$ is easily shown to be $O(r)$ as $r\to 0$, uniformly in $0\leqslant z\leqslant 1$, and hence $O(t)$ on and near the body; therefore, on and near the body,

$$\phi=\frac{U}{2\pi}\left\{S'(z)\log\left(\tfrac{1}{2}\beta r\right)+\frac{1}{2}\int_z^1 S''(s)\log(s-z)\,ds\right.$$
$$\left.-\frac{1}{2}\int_0^z S''(s)\log(z-s)\,ds\right\}+O(t^3). \quad (9.1.17)$$

In a similar way, it can be shown that, on and near the body,

$$\frac{\partial\phi}{\partial z}=\frac{U}{2\pi}\left\{S''(z)\log\left(\tfrac{1}{2}\beta r\right)+\frac{1}{2}\int_z^{1+0}\log(s-z)\,dS''(s)\right.$$
$$\left.-\frac{1}{2}\int_{-0}^{z}\log(z-s)\,dS''(s)\right\}+O(t^3), \quad (9.1.18)$$

where the integrals are of Stieltjes's form, and

$$S''(1+0)=S''(-0)=0.$$

From (9.1.17) and (9.1.18) it is seen that ϕ and $\partial\phi/\partial z$ are $O(t^2\log t)$ on and near the body, and $\partial^2\phi/\partial z^2$ is also of this order, whereas $\partial\phi/\partial r$ is $O(t)$ and $\partial^2\phi/\partial r^2$ is $O(1)$. It follows that the linearizing approximation used in obtaining the equation for ϕ requires further justification, since some of the terms neglected (e.g. $(\partial\phi/\partial r)^2(\partial^2\phi/\partial r^2)$) are of the same order as $\partial^2\phi/\partial z^2$, which is retained. The expression for ϕ on and near the body, given in (9.1.17), is the slender-body approximation for ϕ in the case under consideration here; the accuracy of this approximation is investigated below, in § 9.4, where it is shown that the error term is actually $O(t^4\log^2 t)$ relative to the exact (inviscid flow) potential.

Near to the axis for $z<0$, $z>1$, it can be shown that ϕ is $O(r^2)$, and $\partial\phi/\partial r$ is $O(r)$, but not uniformly near the pointed ends, as is to be expected, because $\partial\phi/\partial r = UR'(0)$ on the body near $z=0$, for example.

When the pressure is calculated, the quadratic approximation to Bernoulli's equation has to be used in the form

$$(p-p_0)/\rho_0 = -U\,\partial\phi/\partial z - \tfrac{1}{2}(\partial\phi/\partial r)^2, \qquad (9.1.19)$$

the terms neglected being $O(t^4 \log^2 t)$.

Consideration of the flow when the body is placed at a small incidence to the undisturbed stream is made later.

9.2. The linearized potential for axially symmetrical supersonic flow past a body of revolution

The supersonic analogue of the potential (9.1.1) for a body of revolution, whose upstream pointed end is at $z=0$, is

$$\phi = \int_0^{z-Br} \frac{f(s)}{\sqrt{\{(z-s)^2 - B^2r^2\}}}\,ds, \qquad (9.2.1)$$

which is the potential of an axial distribution of supersonic sources in $z>0$, but a little care is required to make sure that this is the required form of solution.

That (9.2.1) does satisfy the linearized potential equation can easily be verified, either by using the calculus of finite parts, or by putting

$$s = z - Br\cosh\lambda, \qquad (9.2.2)$$

in which case (9.2.1) becomes

$$\phi = \int_0^{\cosh^{-1}(z/Br)} f(z - Br\cosh\lambda)\,d\lambda, \qquad (9.2.3)$$

and the differentiations can be carried out directly. This form (9.2.3) also shows that the potential represents disturbances being propagated along the cones $z - Br\cosh\lambda = \textit{constant}$, and so outwards and downstream, as is required.

It follows from the form of the potential that the flow at any point (r_1, z_1), say, is independent of conditions downstream from the cone

$$z - Br = z_1 - Br_1, \qquad (9.2.4)$$

a result which is otherwise evident on physical grounds. Hence, in the case of supersonic motion, the body need not have a pointed downstream end, and may have a circular base provided that a knowledge of the flow in the influence domain of the base is not required. The aerodynamic force on such a body can be calculated if the value of the base pressure is known. As before, the body will be taken to be of unit length, with its pointed nose at $z=0$, and its base in the plane $z=1$.

Let the body surface be given by

$$r=R(z) \quad \text{for} \quad 0\leqslant z\leqslant 1, \tag{9.2.5}$$

where $R(z)$ is a piece-wise analytic function of z which satisfies the conditions

$$R(z)<Atz, \quad |R'(z)|<At, \quad |R''(z)|<At, \quad |R'''(z)|<At, \tag{9.2.6}$$

where t is the thickness ratio of the body, as before. The boundary condition on the body surface is given by (9.1.4), and the boundary condition that the disturbance vanishes upstream from the body is satisfied by the form of the potential,

$$\left.\begin{aligned} \phi&=0 \quad \text{for} \quad z-Br<0,\\ \phi&=\int_0^{z-Br}\frac{f(s)}{\sqrt{\{(z-s)^2-B^2r^2\}}}\,\mathrm{d}s \quad \text{for} \quad 0<z-Br<1,\end{aligned}\right\} \tag{9.2.7}$$

and if the body is pointed at the base, so that $R(z)<Atz(1-z)$,

$$\phi=\int_0^1\frac{f(s)}{\sqrt{\{(z-s)^2-B^2r^2\}}}\,\mathrm{d}s \quad \text{for} \quad z-Br>1. \tag{9.2.8}$$

By methods similar to those used in §9.1, it can be shown that

$$f(z)=-(U/2\pi)\,S'(z)\{1+O(t^2\log t)\}, \tag{9.2.9}$$

and that, on and near the body,

$$\phi=\frac{U}{2\pi}\left\{S'(z)\log\tfrac{1}{2}Br-\int_0^z S''(s)\log(z-s)\,\mathrm{d}s\right\}+O(t^3), \tag{9.2.10}$$

$$\frac{\partial\phi}{\partial z}=\frac{U}{2\pi}\left\{S''(z)\log\tfrac{1}{2}Br-\int_{-0}^z\log(z-s)\,\mathrm{d}S''(s)\right\}+O(t^3). \tag{9.2.11}$$

The remarks made about the solution for subsonic flow, at the end of §9.1, apply equally to this solution for supersonic flow.

The potential for supersonic flow past a slender body of revolution was first obtained by von Karman and Moore (1932), who gave both of the forms (9.2.1) and (9.2.3). The form (9.2.10) (with a different O-term, representing a better approximation; see § 9.4) was given by Lighthill (1945).

9.3. The linearized potentials for flows past more general bodies

In order to determine the flows past slender bodies whose cross-sections are not circles, or past a slender body of revolution at incidence, solutions of the linearized potential equation are required which depend upon the polar angle, θ, in cylindrical polar co-ordinates. If the velocities are to be single-valued functions, this dependence on θ must occur through factors $\cos n\theta$ and $\sin n\theta$, where n is an integer. Such solutions, which are generalizations of the forms used in §§ 9.1 and 9.2, can be constructed, and prove to be suitable for the discussion of the flows past slender pointed bodies.

For subsonic flow, the potential (9.1.1) can be written as

$$\phi = \int_{-\infty}^{\infty} \frac{f(s)}{\sqrt{\{(z-s)^2+\beta^2 r^2\}}}\, ds, \tag{9.3.1}$$

where it is to be understood that $f(z)$ vanishes outside $0 \leqslant z \leqslant 1$ (or outside the body, if it is not of unit length). On making the substitution

$$s = z + \beta r \sinh \lambda, \tag{9.3.2}$$

(9.3.1) becomes

$$\phi = \int_{-\infty}^{\infty} f(z+\beta r \sinh \lambda)\, d\lambda. \tag{9.3.3}$$

The equation for ϕ in cylindrical polar co-ordinates, r, θ, z, is

$$\frac{\partial^2 \phi}{\partial r^2} + \frac{1}{r}\frac{\partial \phi}{\partial r} + \frac{1}{r^2}\frac{\partial^2 \phi}{\partial \theta^2} + (1-M^2)\frac{\partial^2 \phi}{\partial z^2} = 0, \tag{9.3.4}$$

and a generalization of (9.3.3), in the form

$$\phi = {\cos \atop \sin} n\theta \int_{-\infty}^{\infty} f_n(z+\beta r \sinh \lambda)\, g_n(\lambda)\, d\lambda, \tag{9.3.5}$$

can be shown to satisfy (9.3.4), provided that

$$\left[\frac{\beta}{r} f_n' g_n \cosh \lambda - \frac{1}{r^2} f_n g_n'\right]_{-\infty}^{\infty} + \frac{1}{r^2}\int_{-\infty}^{\infty} f_n\{g_n'' - n^2 g_n\}\, d\lambda = 0. \tag{9.3.6}$$

This last condition is satisfied by

$$g_n(\lambda) = e^{-n\lambda}, \tag{9.3.7}$$

provided that

$$\left.\begin{aligned} &f_n = 0 \quad \text{and} \quad f_n' = 0 \quad \text{for} \quad \lambda < \lambda_0 < \infty, \\ \text{and} \quad &f_n' = o(1) \quad \text{as} \quad \lambda \to +\infty. \end{aligned}\right\} \tag{9.3.8}$$

On transforming back to the form (9.3.1) by using (9.3.2), (9.3.5) becomes

$$\phi = \frac{\cos}{\sin} n\theta \int_{-\infty}^{\infty} \frac{(\sqrt{\{(z-s)^2+\beta^2r^2\}}+z-s)^n f_n(s)}{\beta^n r^n \sqrt{\{(z-s)^2+\beta^2r^2\}}} ds. \tag{9.3.9}$$

For small r,

$$\begin{aligned} \frac{1}{\beta r}(\sqrt{\{(z-s)^2+\beta^2r^2\}}+z-s) \\ &= \frac{2(z-s)}{\beta r} + O(r) \quad \text{for} \quad z > s, \\ &= \frac{\beta r}{2(z-s)} + O(r^3) \quad \text{for} \quad z < s, \end{aligned} \tag{9.3.10}$$

but not uniformly near $z = s$. Thus the fundamental solution,

$$\phi = \frac{\cos}{\sin} n\theta \frac{(\sqrt{\{(z-s)^2+\beta^2r^2\}}+z-s)^n}{r^n\sqrt{\{(z-s)^2+\beta^2r^2\}}}, \tag{9.3.11}$$

is singular on the axis for $z > s$, and vanishes on the axis for $z < s$. When $n = 1$, (9.3.11) gives

$$\phi = \frac{\cos}{\sin} \theta \frac{1}{r}\left(1 + \frac{z-s}{\sqrt{\{(z-s)^2+\beta^2r^2\}}}\right), \tag{9.3.12}$$

which is the potential of a vortex pair (cf. §4.13) on $r = 0$, $z > s$. The potentials given by (9.3.11) may thus be considered as being due to 'vortex multiplets' on $r = 0$, $z > s$, and the potential (9.3.9) represents a distribution of such 'vortex multiplets'. The case $n = 0$ is that investigated in §9.1, and represents a distribution of sources.

For the case of a pointed body with its nose at $z = 0$, it is evident from (9.3.10) that

$$f_n(z) = 0 \quad \text{for} \quad z < 0, \tag{9.3.13}$$

since the perturbation velocity must be finite on the z-axis for $z < 0$, and must vanish as $z \to -\infty$.

On and near the body surface, where r is small, (9.3.9) can be shown to give

$$\phi = \frac{\cos}{\sin} n\theta \frac{2^n}{\beta^n r^n} \int_0^z (z-s)^{n-1} f_n(s)\, ds\{1 + O(t)\} \tag{9.3.14}$$

when $n \geqslant 1$. Hence the general expression for the potential at such points can be written

$$\phi_0 = a_0(z)\log r + b_0(z) + \sum_{n=1}^{\infty} \frac{a_n(z)}{r^n}\cos(n\theta+\theta_n),\dagger \quad (9.3.15)$$

where the a_n's and θ_n's are functions of z to be determined from the boundary condition on the body, and $b_0(z)$ is given from (9.1.17) as

$$b_0(z) = a_0(z)\log\tfrac{1}{2}\beta + \frac{1}{2}\int_z^1 a_0'(s)\log(s-z)\,ds - \frac{1}{2}\int_0^z a_0'(s)\log(z-s)\,ds, \quad (9.3.16)$$

for a body of unit length. The error in (9.3.15) is a factor $1+O(t)$.

A similar generalization can be made for supersonic flow. The formula (9.2.3) can be written

$$\phi = \int_0^\infty f(z - Br\cosh\lambda)\,d\lambda, \quad (9.3.17)$$

and this can be generalized to

$$\phi = \frac{\cos}{\sin} n\theta \int_0^\infty f_n(z-Br\cosh\lambda)\,h_n(\lambda)\,d\lambda, \quad (9.3.18)$$

which satisfies the linearized potential equation if

$$\left[-\frac{1}{r^2}f_n h_n'\right]_0^\infty + \frac{1}{r^2}\int_0^\infty f_n\{h_n'' - n^2 h_n\}\,d\lambda = 0. \quad (9.3.19)$$

This last condition is satisfied if

$$h_n'' - n^2 h_n = 0 \quad \text{and} \quad h_n'(0) = 0; \quad (9.3.20)$$

that is, if

$$h_n(\lambda) = \cosh n\lambda. \quad (9.3.21)$$

On transforming back to the form (9.2.1) by the transformation (9.2.2), (9.3.18) with (9.3.21) gives, for a body whose pointed end is at $z=0$,

$$\phi = \frac{\cos}{\sin} n\theta \int_0^{z-Br} \frac{1}{2}\left\{\left[\frac{z-s+\sqrt{\{(z-s)^2-B^2r^2\}}}{Br}\right]^n + \left[\frac{z-s-\sqrt{\{(z-s)^2-B^2r^2\}}}{Br}\right]^n\right\} \frac{f_n(s)}{\sqrt{\{(z-s)^2-B^2r^2\}}}\,ds \quad (9.3.22)$$

when $z - Br > 0$, and $\phi = 0$ for $z - Br < 0$.

† It is convenient to take $\phi = U\phi_0$, so that $\mathbf{v} = U\nabla\phi_0$; the suffix zero also serves to distinguish this potential function from the true linearized potential.

On and near the body surface, where r is small, (9.3.22) can be shown to give

$$\phi = {\cos \atop \sin} n\theta \frac{2^n}{B^n r^n} \int_0^z (z-s)^{n-1} f_n(s)\,ds\{1+O(r)\}, \quad (9.3.23)$$

when $n \geqslant 1$. Hence the expression for the potential at points on and near the body surface is formally the same as (9.3.15), but in the supersonic case $b_0(z)$ is given from (9.2.10) by

$$b_0(z) = a_0(z) \log \tfrac{1}{2}B - \int_0^z a_0'(s) \log (z-s)\,ds.\dagger \quad (9.3.24)$$

By writing $\zeta = x + iy = re^{i\theta}$, (9.3.15) for both subsonic and supersonic flow may be expressed as the real part of

$$w_0 = \phi_0 + i\psi_0 = a_0(z) \log \zeta + b_0(z) + \sum_{n=1}^{\infty} a_n(z) \zeta^{-n}, \quad (9.3.25)$$

where ψ_0 is a real function of r, θ, z, which is not a true stream function for the motion, but is denoted by ψ_0 because it has the properties of a stream function in a two-dimensional incompressible flow. Thus ϕ_0 is a harmonic function of r, θ, satisfying the equation

$$\frac{\partial^2 \phi_0}{\partial x^2} + \frac{\partial^2 \phi_0}{\partial y^2} = 0, \quad (9.3.26)$$

and the results for two-dimensional incompressible flow can be used to determine $w_0 - b_0$ from the boundary conditions on the body. The function b_0 can then be determined from a_0 by using (9.3.16) for subsonic flow, or (9.3.24) for supersonic flow. The form of (9.3.25) shows that *the incompressible flow must have vanishing velocity at* $|\zeta| = \infty$.

If the slender body is contained inside a cylinder $r = r_1$, then it is apparent that the series (9.3.25) converges for $|\zeta| \geqslant r_1$, and may be continued analytically outside the body for $|\zeta| < r_1$ if it does not converge in this region. Usually, however, $w_0 - b_0$ is obtained as a closed expression, and a_0, a_1, etc., have to be obtained by expanding this expression in a series of the form (9.3.25) (see §§ 9.9, 9.10 and 9.11).

The equation (9.3.26) can be derived from the linearized potential equation in both subsonic and supersonic flow by assuming

† These expressions for the potential in the slender-body approximation can also be derived by operational methods, as has been done by Adams and Sears (1953) for subsonic flow, and by Ward (1949*a*) for supersonic flow.

that the variations in the flows as x and y vary are large compared with the variations with z, which is a reasonable assumption for slender bodies. This approach was used by Munk (1924) in his theory of the forces on airship hulls in incompressible flow, and by R. T. Jones (1946*a*) for a treatment of wings of low aspect-ratio in compressible flows, but the method is not satisfactory by itself, since it does not give the term b_0.

9.4. The accuracy of the slender-body approximation

With certain restrictions on the shape of the slender body, the accuracy of the solution on and near the body surface, given by the slender-body approximation to the potential, ϕ_0, can be determined by calculating the order of the next approximation to the exact (inviscid flow) potential. The body, either pointed at both ends, or with a flat base in supersonic flow, is taken to be of unit length and of maximum thickness t. For a slender body, t must be small compared with unity; the angle which any tangent plane to the body surface makes with the undisturbed stream direction must be small and $O(t)$, and the rate of change of this angle with z must also be small and $O(t)$. (This last restriction is not entirely necessary in supersonic flow, and is relaxed in §§9.12 and 9.13 for bodies of revolution with discontinuous meridian profiles.) To ensure that the perturbation velocity is everywhere small, a further condition on the shape of the body is required; this concerns the radius of curvature of any section of the body surface by a plane $z = constant$. If d is the maximum diameter of the section, the curvature must be $O(1/d)$ for all points where the section is convex outwards; there is no restriction at points where the section is concave outwards. This condition is not always necessary for bodies at zero incidence, but is always required if the flow at incidence is to be calculated within a known approximation.

Some interesting body shapes are excluded by the above curvature condition, and it may be relaxed under some circumstances, but then the accuracy of the solution becomes difficult to estimate. Examples of such bodies are considered in §§9.10 and 9.11.

In determining the accuracy of ϕ_0 as a solution of the full non-linear potential equation, it is sufficient to consider only the equation for a perfect gas with constant specific heats, (1.8.3), and it is convenient to express this equation in terms of independent variables $\zeta, \bar{\zeta}$, where $\zeta = x + iy$, $\bar{\zeta} = x - iy$. With these variables, (1.8.3) becomes

$$c^2\left(4\frac{\partial^2\Phi}{\partial\zeta\partial\bar{\zeta}} + \frac{\partial^2\Phi}{\partial z^2}\right) = 4\left(\frac{\partial\Phi}{\partial\zeta}\right)^2\frac{\partial^2\Phi}{\partial\bar{\zeta}^2} + 4\frac{\partial\Phi}{\partial\zeta}\frac{\partial\Phi}{\partial\bar{\zeta}}\frac{\partial^2\Phi}{\partial\zeta\partial\bar{\zeta}}$$

$$+4\left(\frac{\partial\Phi}{\partial\bar{\zeta}}\right)^2\frac{\partial^2\Phi}{\partial\zeta^2} + \frac{\partial\Phi}{\partial z}\frac{\partial}{\partial z}\left(\frac{\partial\Phi}{\partial\zeta}\frac{\partial\Phi}{\partial\bar{\zeta}}\right) + \left(\frac{\partial\Phi}{\partial z}\right)^2\frac{\partial^2\Phi}{\partial z^2}, \quad (9.4.1)$$

where $$c^2 = c_0^2 - \tfrac{1}{2}(\gamma - 1)\left\{4\frac{\partial\Phi}{\partial\zeta}\frac{\partial\Phi}{\partial\bar{\zeta}} + \left(\frac{\partial\Phi}{\partial z}\right)^2 - U^2\right\}. \quad (9.4.2)$$

It is necessary at this stage to assume some results that are proved later; when the coefficients in ϕ_0 are determined, it is found that a_0 and b_0 are $O(t^2)$ and that the a_n's ($n = 1, 2, 3, \ldots$) are $O(t^{n+2})$. By using these results it is seen that, near the body, where the series converges,

$$\frac{\partial\phi_0}{\partial z},\ \frac{\partial^2\phi_0}{\partial z^2},\ \text{etc., are } O(t^2\log t);\quad \frac{\partial\phi_0}{\partial\zeta},\ \frac{\partial^2\phi_0}{\partial\zeta\partial z},\ \text{etc., are } O(t);$$

and $$\frac{\partial^2\phi_0}{\partial\zeta^2},\ \text{etc., are } O(1). \quad (9.4.3)$$

Now assume that

$$\Phi = U(z + \phi_0 + \phi_1) = U\{z + \tfrac{1}{2}(w_0 + \bar{w}_0) + \phi_1\}, \quad (9.4.4)$$

where ϕ_1 and its derivatives are of sufficiently smaller order than ϕ_0 and its corresponding derivatives so that squares and products of ϕ_1 and its derivatives can be neglected in (9.4.1). Then (9.4.1) gives

$$4\frac{\partial^2\phi_1}{\partial\zeta\partial\bar{\zeta}} + (1 - M^2)\frac{\partial^2\phi_1}{\partial z^2} = M^2\left\{\frac{\partial}{\partial z}\left(\frac{\partial w_0}{\partial\zeta}\frac{\partial\bar{w}_0}{\partial\bar{\zeta}}\right) + \frac{1}{2}\left(\frac{\partial w_0}{\partial\zeta}\right)^2\frac{\partial^2\bar{w}_0}{\partial\bar{\zeta}^2}\right.$$

$$\left. + \frac{1}{2}\left(\frac{\partial\bar{w}_0}{\partial\bar{\zeta}}\right)^2\frac{\partial^2 w_0}{\partial\zeta^2}\right\} + \tfrac{1}{2}(M^2 - 1)\left(\frac{\partial^2 w_0}{\partial z^2} + \frac{\partial^2\bar{w}_0}{\partial z^2}\right) + O(t^4\log^2 t)$$

$$= f(\zeta, \bar{\zeta}, z), \text{ say.} \quad (9.4.5)$$

A particular integral can be obtained, by solving in series, as

$$\phi_1 = \frac{1}{4}\int^{\zeta}\int^{\bar{\zeta}} f(\zeta, \bar{\zeta}, z)\,d\zeta\,d\bar{\zeta}$$

$$+ \tfrac{1}{16}(M^2 - 1)\frac{\partial^2}{\partial z^2}\int^{\zeta}\int^{\zeta}\int^{\bar{\zeta}}\int^{\bar{\zeta}} f(\zeta, \bar{\zeta}, z)\,(d\zeta)^2(\partial\bar{\zeta})^2 + \ldots$$

$$= \tfrac{1}{8}M^2\left\{2\frac{\partial}{\partial z}(w_0\bar{w}_0) + \frac{\partial\bar{w}_0}{\partial\bar{\zeta}}\int^{\zeta}\left(\frac{\partial w_0}{\partial\zeta}\right)^2 d\zeta + \frac{\partial w_0}{\partial\zeta}\int^{\bar{\zeta}}\left(\frac{\partial\bar{w}_0}{\partial\bar{\zeta}}\right)^2 d\bar{\zeta}\right\}$$

$$+ \tfrac{1}{8}(M^2 - 1)\frac{\partial^2}{\partial z^2}\left(\bar{\zeta}\int^{\zeta} w_0\,d\zeta + \zeta\int^{\bar{\zeta}} \bar{w}_0\,d\bar{\zeta}\right) + \ldots, \quad (9.4.6)$$

from which it is seen that

$$\left.\begin{array}{lll}\phi_1,\ \partial\phi_1/\partial z,\ \partial^2\phi_1/\partial z^2,\ \text{etc.}, & \text{are} & O(t^4\log^2 t),\\ \partial\phi_1/\partial\zeta,\ \partial^2\phi_1/\partial\zeta\,\partial z,\ \text{etc.}, & \text{are} & O(t^3\log t),\\ \partial^2\phi_1/\partial\zeta^2,\ \text{etc.}, & \text{are} & O(t^2\log t).\end{array}\right\} \quad (9.4.7)$$

The terms omitted in (9.4.6) are of smaller order than those shown.

The conditions on the shape of the body ensure that ϕ_0 and its derivatives are of the same order of magnitude on the body as they are near it, so the order of magnitude of ϕ_1 and its derivatives on and near the body are those given above. Hence ϕ_0 gives the perturbation velocity on and near the body within a factor $1+O(t^2\log t)$ on the assumption, implicit in the above analysis, that M is not large. For large M this factor takes the form $1+O(M^2t^2\log Mt)$, as can easily be shown. When M is nearly unity, the slender-body approximation fails owing to the occurrence of the terms in $\log\beta$ or $\log B$ in the coefficient b_0; thus the method fails to give accurate results in the transonic range. The entropy changes at the bow shock in supersonic flow can be shown to be insignificant so far as the present theory is concerned; the error from this cause is $O(M^{12}t^{12})$.

9.5. The boundary condition on the body

Let the contour C be the cross-section of the body surface by a plane $z=constant$, and let C' represent the projection on the same plane of the cross-section at $z+\mathrm{d}z$, as in fig. 9.1. If ν is distance along the outward normals to C, τ is distance along C, and $\mathrm{d}\nu$ is the distance between C and C', then the boundary condition on the body surface is

$$\frac{\partial}{\partial\nu}(\phi_0+\phi_1)=\frac{\mathrm{d}\nu}{\mathrm{d}z}\left\{1+\frac{\partial}{\partial z}(\phi_0+\phi_1)\right\}. \quad (9.5.1)$$

Now $\mathrm{d}\nu/\mathrm{d}z$ is $O(t)$ by definition, so $\partial\phi_0/\partial\nu$ must be $O(t)$ on the body, and if the curvature of C satisfies the condition of §9.4, then $\partial\phi_0/\partial\tau$ is also $O(t)$ on the body. Thus $\partial w_0/\partial\zeta$ is $O(t)$ on the body and near it, and the a_n's are $O(t^{n+2})$ $(n=0, 1, 2, \ldots)$. $\partial\phi_0/\partial z$ is then $O(t^2\log t)$, and the derivatives of ϕ_1 have the orders given in (9.4.7). Hence (9.5.1) becomes

$$\frac{\partial\phi_0}{\partial\nu}=\frac{\mathrm{d}\nu}{\mathrm{d}z}+O(t^3\log t), \quad (9.5.2)$$

and it follows from (9.4.3) and (9.4.7) that, if ϕ_0 is determined by the condition

$$\frac{\partial \phi_0}{\partial \nu} = \frac{d\nu}{dz}, \tag{9.5.3}$$

then the velocity components are given by

$$\left.\begin{aligned} u &= U\frac{\partial \phi_0}{\partial x} + O(t^3 \log t), \\ v &= U\frac{\partial \phi_0}{\partial y} + O(t^3 \log t), \\ w &= U\left(1 + \frac{\partial \phi_0}{\partial z}\right) + O(t^4 \log^2 t). \end{aligned}\right\} \tag{9.5.4}$$

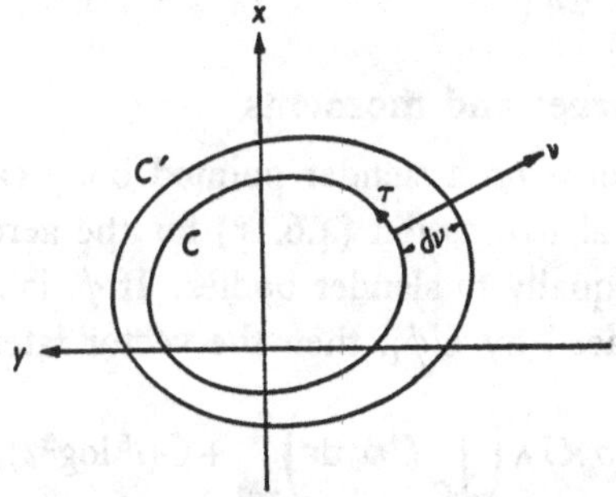

Fig. 9.1. Sections of a slender body by two planes perpendicular to the stream direction and separated by a distance dz.

9.6 The coefficients a_0 and b_0

Consider the integral

$$\int_C \frac{\partial \phi_0}{\partial \nu} d\tau. \tag{9.6.1}$$

By using Gauss's theorem, since ϕ_0 satisfies Laplace's equation in two dimensions, (9.3.26), the contour C may be replaced by a circle of small radius r_1, with its centre on the z-axis, on which the series for ϕ_0 converges absolutely. Hence, by using (9.3.15) and integrating term by term

$$\int_C \frac{\partial \phi_0}{\partial \nu} d\tau = \int_0^{2\pi} \left(\frac{\partial \phi_0}{\partial r}\right)_{r=r_1} r_1 d\theta = 2\pi a_0(z). \tag{9.6.2}$$

Also, from the boundary condition (9.5.3), if the area enclosed by C is denoted by $S(z)$,

$$\int_C \frac{\partial \phi_0}{\partial \nu} d\tau = \int_C \frac{d\nu}{dz} d\tau = S'(z). \tag{9.6.3}$$

Thus, by comparing (9.6.2) and (9.6.3),

$$a_0(z) = S'(z)/2\pi, \tag{9.6.4}$$

and, for subsonic flow, from (9.3.16), $b_0(z)$ is given by

$$b_0(z) = \frac{1}{2\pi}\left\{S'(z)\log\tfrac{1}{2}\beta + \frac{1}{2}\int_z^1 S''(s)\log(s-z)\,ds - \frac{1}{2}\int_0^z S''(s)\log(z-s)\,ds\right\}, \tag{9.6.5}$$

and, for supersonic flow, from (9.3.24),

$$b_0(z) = \frac{1}{2\pi}\left\{S'(z)\log\tfrac{1}{2}B - \int_0^z S''(s)\log(z-s)\,ds\right\}. \tag{9.6.6}$$

9.7. Lateral forces and moments

The lateral force on a slender pointed body can be calculated from the general expression (4.6.15) for the aerodynamic force, which applies equally to slender bodies. If ϕ_{I} is replaced by $U\phi_0$ and ϕ_{II} is replaced by $U\phi_1$, then the vector lateral force is seen to be

$$\rho_0 \mathbf{U} \wedge \left(\int_C U\phi_0\,d\mathbf{s}\right)_{z=1} + O(t^5\log^2 t), \tag{9.7.1}$$

the integral being taken over the base section of the body. It is more convenient in this case to consider the complex force $F = F_x + iF_y$, where F_x, F_y are the components of the lateral force parallel to the x- and y-axes respectively. Then, from (9.7.1),

$$F = -\rho_0 U^2\left(\int_C \phi_0\,d\zeta\right)_{z=1} + O(t^5\log^2 t). \tag{9.7.2}$$

This expression for F can be put into a more convenient form as follows. Since $\phi_0 = w_0 - i\psi_0$,

$$\int_C \phi_0\,d\zeta = \int_C w_0\,d\zeta - i\int_C \psi_0\,d\zeta. \tag{9.7.3}$$

Now w_0 is an analytic function of ζ in a cut ζ-plane, and C encloses all the singularities of w_0, none of which lies on C. Thus the series for ϕ_0 converges on some arc of C, and another contour C_1 can be chosen, on the whole of which the series converges, and which has at least one common point, ζ_0 say, with C. The cut in the ζ-plane can be chosen to pass through ζ_0, and if C and C_1 start and finish at ζ_0 (on opposite sides of the cut), then

$$\int_C w_0\,d\zeta = \int_{C_1} w_0\,d\zeta = \int_{C_1}\left(\frac{1}{2\pi}S'(z)\log\zeta + b_0 + \frac{a_1}{\zeta} + \dots\right)d\zeta$$

$$= \frac{1}{2\pi}S'(z)\,2\pi i\zeta_0 + 2\pi i a_1. \tag{9.7.4}$$

Also, since $\psi_0 = \frac{1}{2\pi}S'(z)\,\theta + \text{single-valued function}$, (9.7.5)

and
$$\frac{\partial\psi_0}{\partial\tau} = \frac{\partial\phi_0}{\partial\nu} = \frac{d\nu}{dz} \text{ on } C, \tag{9.7.6}$$

then
$$\int_C \psi_0\,d\zeta = [\psi_0\zeta]_C - \int_C \zeta\frac{\partial\psi_0}{\partial\tau}\,d\tau$$

$$= S'(z)\,\zeta_0 - \int_C \zeta\frac{d\nu}{dz}\,d\tau. \tag{9.7.7}$$

The last integral has a simple geometrical interpretation; it is the rate of change (with respect to z) of the first moment of the area $S(z)$ enclosed by C. If the co-ordinates of the centre of area of $S(z)$ are $x_g(z)$, $y_g(z)$, and if

$$\zeta_g(z) = x_g(z) + iy_g(z), \tag{9.7.8}$$

then
$$\int_C \zeta\frac{d\nu}{dz}\,d\tau = \frac{d}{dz}\{S(z)\,\zeta_g(z)\} = S'(z)\,\zeta_g(z) + S(z)\,\zeta_g'(z). \tag{9.7.9}$$

By substituting all these results in (9.7.2), the final expression for F is

$$F/\rho_0 U^2 = 2\pi a_1(1) + S'(1)\,\zeta_g(1) + S(1)\,\zeta_g'(1). \tag{9.7.10}$$

The moments of the lateral forces about the x- and y-axes can be calculated by observing that $F(z)$, defined by

$$F(z)/\rho_0 U^2 = 2\pi a_1(z) + S'(z)\,\zeta_g(z) + S(z)\,\zeta_g'(z), \tag{9.7.11}$$

determines the lateral force on that part of the body which lies upstream from the plane z, and that the complex lateral force per unit length is $F'(z)$. Thus the complex moment, M_ζ, is

$$M_\zeta = M_x + iM_y = i\int_0^1 zF'(z)\,dz = iF(1) - i\int_0^1 F(z)\,dz, \tag{9.7.12}$$

and on substituting the value of $F(z)$ from (9.7.11),

$$M_\zeta/\rho_0 U^2 = 2\pi i a_1(1) - 2\pi i\int_0^1 a_1(z)\,dz$$
$$+ i\{S'(1)\,\zeta_g(1) + S(1)\,\zeta_g'(1) - S(1)\,\zeta_g(1)\} + O(t^5\log^2 t). \tag{9.7.13}$$

The above results are given in their most general form for a slender body satisfying the conditions given in § 9.4. Some special cases are given below. It will be noticed that $b_0(z)$ does not occur in these expressions, so they apply to both subsonic and supersonic flow. The coefficient $a_1(z)$, on which the lateral forces and moments depend, is determined entirely by the flow in an infinitesimal neighbourhood of the section z; so even in the case of subsonic flow, the lateral forces and moments are independent of the form of the wake, and the body can have a flat base. Of course the complete flow pattern cannot be determined since the form of the wake would affect $b_0(z)$ (but this does not matter provided that the pressure distribution is not required); in particular, the drag force cannot be calculated.

If the body is pointed at both ends, $a_1(1)=0$, and (9.7.10) shows that the lateral force vanishes to $O(t^5 \log^2 t)$. The moment does not vanish in general.

9.8. The drag force

The drag force on a slender body can be calculated from the general result (4.6.15), as is done above for the lateral forces, but in this case it is easier to determine the order of the approximation by proceeding directly. Just as in § 4.6, the drag force can be calculated from the net momentum flux in the z-direction through a closed surface which completely encloses the body and passes through the trailing edge, which in this case is C for $z=1$. This surface is taken to be a cylinder $r=r_1$, which contains the body completely, with plane ends at $z=0$ and $z=1$. Let S_1 denote the end at $z=0$, S_2 denote the curved surface $r=r_1$, and S_3 denote that part of the end at $z=1$ which is outside the body. Then the drag force, D, parallel to the z-axis, is given by

$$D=\int_{S_1}\left\{p+\rho\left(\frac{\partial\Phi}{\partial z}\right)^2\right\}dS-\int_{S_2}\rho\frac{\partial\Phi}{\partial r}\frac{\partial\Phi}{\partial z}\,dS$$

$$-\int_{S_3}\left\{p+\rho\left(\frac{\partial\Phi}{\partial z}\right)^2\right\}dS-p_B S(1), \quad (9.8.1)$$

where p_B is the base pressure if $S(1)\neq 0$.

Now from the previous results on orders of magnitude,

$$\frac{p-p_0}{\rho_0 U^2}=-\frac{\partial\phi_0}{\partial z}-\frac{1}{2}\left(\frac{\partial\phi_0}{\partial r}\right)^2-\frac{1}{2r^2}\left(\frac{\partial\phi_0}{\partial\theta}\right)^2+O(t^4\log^2 t), \quad (9.8.2)$$

and $$\rho=\rho_0\{1+O(t^2\log t)\}. \tag{9.8.3}$$

Hence, on substitution of these last two results, (9.8.1) can be shown to give

$$\frac{D}{\frac{1}{2}\rho_0 U^2}=-2\int_{S_3}\frac{\partial\phi_0}{\partial r}\frac{\partial\phi_0}{\partial z}\,dS+\int_{S_2}\left\{\left(\frac{\partial\phi_0}{\partial r}\right)^2+\frac{1}{r^2}\left(\frac{\partial\phi_0}{\partial\theta}\right)^2\right\}dS$$
$$+\frac{p_0-p_B}{\frac{1}{2}\rho_0 U^2}S(1)+O(t^6\log^2 t). \tag{9.8.4}$$

There is no integral over S_1 because ϕ_0 vanishes there for a pointed body. The quantity $(p_0-p_B)/\frac{1}{2}\rho_0 U^2$ is the base-pressure coefficient and is denoted by C_{pB}.

The integral over S_3 can be transformed into integrals taken round the boundaries of S_3, by using Green's theorem, and (9.8.4) then becomes

$$\frac{D}{\frac{1}{2}\rho_0 U^2}=-2\int_0^1\int_0^{2\pi}\left(\frac{\partial\phi_0}{\partial r}\frac{\partial\phi_0}{\partial z}\right)_{r=r_1}r_1\,dr\,d\theta$$
$$+\left(\int_0^{2\pi}\left(\phi_0\frac{\partial\phi_0}{\partial r}\right)_{r=r_1}r_1\,d\theta\right)_{z=1}$$
$$-\left(\int_C\phi_0\frac{\partial\phi_0}{\partial\nu}\,d\tau\right)_{z=1}+C_{pB}S(1)+O(t^6\log^2 t). \tag{9.8.5}$$

The series for ϕ_0 can now be substituted in the first two integrals, and it can be shown that nearly all the terms cancel, and (9.8.5) gives

$$\frac{D}{\frac{1}{2}\rho_0 U^2}=4\pi\int_0^1 a_0'(z)\,b_0(z)\,dz-2\pi a_0(1)\,b_0(1)$$
$$-\left(\int_C\phi_0\frac{\partial\phi_0}{\partial\nu}\,d\tau\right)_{z=1}+C_{pB}S(1)+O(t^6\log^2 t). \tag{9.8.6}$$

For subsonic flow, the body must be pointed at both ends if this result is to apply, since the coefficient b_0 occurs. In this case, on putting in the value of b_0 given by (9.6.5), it is not difficult to show that D vanishes to $O(t^6\log^2 t)$.

For supersonic flow, the body can have a flat base, and on substituting for b_0 from (9.6.6), (9.8.6) gives

$$\frac{D}{\frac{1}{2}\rho_0 U^2}=\frac{1}{2\pi}\int_0^1\int_0^1\log\frac{1}{|z-s|}S''(z)\,S''(s)\,ds\,dz$$
$$-\frac{1}{2\pi}S'(1)\int_0^1\log\frac{1}{1-z}S''(z)\,dz$$
$$-\left(\int_C\phi_0\frac{\partial\phi_0}{\partial\nu}\,d\tau\right)_{z=1}+C_{pB}S(1)+O(t^6\log^2 t). \tag{9.8.7}$$

This appears to be the simplest form into which the expression for the drag can be put for the completely general case. Two special cases arise for which the result assumes a simpler form. These are (i) when $S(1)=0$, and the body is pointed at both ends, and (ii) when the body is of cylindrical form near $z=1$, so that $S'(1)=0$, and the generators of the cylinder are parallel to the z-axis, which makes $\partial\phi_0/\partial\nu = d\nu/dz = 0$ (the body may be thought of as being at zero incidence in this case). For both (i) and (ii), omitting the base drag term for (ii),

$$\frac{D}{\frac{1}{2}\rho_0 U^2} = \frac{1}{2\pi}\int_0^1\int_0^1 \log\frac{1}{|z-s|} S''(z)\, S''(s)\, ds\, dz + O(t^6 \log^2 t). \quad (9.8.8)$$

This result was obtained by von Karman (1935) and Lighthill (1945) for the special case of a body of revolution at zero incidence and having $S'(1)=0$. It is interesting to compare this result with Prandtl's expression for the induced drag of a lifting line in incompressible flow, or its generalization given by (5.2.14), when it will be seen that the integrals are of the same formal type. In lifting line theory, it is a known result that for a given lift, the induced drag is a minimum for an elliptic lift distribution along the span; thus it follows from the comparison that, for a body of class (ii) with given base area and length, the minimum drag occurs for an elliptic distributon of $S'(z)$. Such a body does not have a pointed nose however, and is therefore not admissible on the present theory. It is, nevertheless, interesting to note that experimental tests in wind tunnels show that bodies with a slight roundness at the nose have lower drags than bodies with pointed noses at supersonic speeds.

The integral around C at $z=1$ in (9.8.7) has a simple interpretation in terms of the incompressible flow associated with the base section; it is twice the kinetic energy per unit length for the movement of the cylinder C in the two-dimensional flow (for density $=1$), a fact which considerably simplifies the calculations in cases for which this quantity is known.

An expression for the variation of drag with incidence can be obtained from (9.8.7). Let the position of zero incidence be defined as the position of zero lateral force, and let ϕ_{00} be the potential in the plane $z=1$ for this position, so that, from (9.7.2),

$$\int_C \phi_{00}\, d\zeta = 0. \quad (9.8.9)$$

Let ϕ_{01} be the extra potential in the plane $z=1$ due to incidence α; then the lateral force is given by

$$\frac{F_x+iF_y}{\rho_0 U^2}=-i\int_C \phi_{01}\,d\zeta=\int_C \phi_{01}\,dy-i\int_C \phi_{01}\,dx. \quad (9.8.10)$$

For the work of this chapter it is convenient to define the incidence α to be a rotation about the y-axis in the negative sense (no loss of generality is incurred, since the direction of the y-axis is not otherwise defined). The boundary condition at C then gives

$$\frac{\partial \phi_{01}}{\partial \nu}=-\alpha \sin \eta, \quad (9.8.11)$$

where η is the angle between the tangent to C at any point and the x-axis.

If D_0 is the drag force at zero incidence, then from (9.8.7),

$$\frac{D-D_0}{\frac{1}{2}\rho_0 U^2}=-\int_C \phi_{00}\frac{\partial \phi_{01}}{\partial \nu}\,d\tau-\int_C \phi_{01}\frac{\partial \phi_{00}}{\partial \nu}\,d\tau$$
$$-\int_C \phi_{01}\frac{\partial \phi_{01}}{\partial \nu}\,d\tau+O(t^6\log^2 t). \quad (9.8.12)$$

These three integrals can be evaluated; for the first,

$$\int_C \phi_{00}\frac{\partial \phi_{01}}{\partial \nu}\,d\tau=-\alpha\int_C \phi_{00}\sin\eta\,d\tau=-\alpha\int_C \phi_{00}\,dy=0 \quad (9.8.13)$$

by using (9.8.11) and (9.8.9); for the second

$$\int_C \phi_{01}\frac{\partial \phi_{00}}{\partial \nu}\,d\tau=\int_C\left(\phi_{01}\frac{\partial \phi_{00}}{\partial \nu}-\phi_{00}\frac{\partial \phi_{01}}{\partial \nu}\right)d\tau$$
$$=\int_0^{2\pi}\left(\phi_{01}\frac{\partial \phi_{00}}{\partial r}-\phi_{00}\frac{\partial \phi_{01}}{\partial r}\right)_{r=r_1} r_1\,d\theta, \quad (9.8.14)$$

from (9.8.13) and by applying Green's theorem. Since the series for ϕ_{00} and ϕ_{01} start with multiples of $\log r$ and $1/r$ respectively, by letting $r_1\to\infty$, it can be shown that this integral vanishes. Finally

$$\int_C \phi_{01}\frac{\partial \phi_{01}}{\partial \nu}\,d\tau=-\alpha\int_C \phi_{01}\sin\eta\,d\tau=-\alpha\int_C \phi_{01}\,dy=-\alpha F_x/\rho_0 U^2 \quad (9.8.15)$$

from (9.8.11) and (9.8.10).

Thus from (9.8.12) and the last three results,

$$D=D_0+\tfrac{1}{2}\alpha F_x+O(t^6\log^2 t). \quad (9.8.16)$$

If F_x is the lift force (in the plane of incidence), (9.8.16) can be written in terms of force coefficients as

$$C_D = C_{D0} + \tfrac{1}{2}(\partial C_L/\partial\alpha)_0\,\alpha^2 + O(t^6 \log^2 t), \qquad (9.8.17)$$

where suffix 0 denotes the values at zero incidence.

9.9. Application to bodies of revolution

Consider a body of revolution at incidence α; then $\zeta_g(z) = -\alpha z$, and the appropriate complex potential w_0 is

$$w_0 = b_0 + \frac{1}{2\pi} S'(z) \log(\zeta - \zeta_g) - \frac{1}{\pi} S(z) \frac{\zeta_g'}{\zeta - \zeta_g}, \qquad (9.9.1)$$

neglecting a factor $1 + O(\alpha^2)$ (since the sections have been taken to be circles, whereas they are really ellipses). $w_0 - b_0$ is composed of a term representing radial flow outwards from ζ_g, and a term due to the cross-motion of the circular section, centre ζ_g. This can be expanded to give

$$w_0 - b_0 = \frac{1}{2\pi} S'(z) \log \zeta - \frac{\zeta_g(z)\, S'(z) + 2\zeta_g'(z)\, S(z)}{2\pi r} + \ldots, \qquad (9.9.2)$$

so the coefficient $a_1(z)$ is

$$a_1(z) = -\frac{1}{2\pi}\{\zeta_g(z)\, S'(z) + 2\zeta_g'(z)\, S(z)\}. \qquad (9.9.3)$$

Therefore, from (9.7.10), if $\alpha = O(t)$, the lateral force F is given by

$$F/\rho_0 U^2 = -\zeta_g'(1)\, S(1) + O(t^5 \log^2 t) = \alpha S(1) + O(t^5 \log^2 t). \qquad (9.9.4)$$

Hence F is real, and the lateral force is a lift force in the direction of the x-axis. If $S(1) = 0$, then there is no lateral force. If $S(1) \neq 0$, then the lift coefficient with respect to base area is

$$C_L = \frac{F_x}{\tfrac{1}{2}\rho_0 U^2 S(1)} = 2\alpha + O(t^3 \log^2 t). \qquad (9.9.5)$$

The moment about the nose is obtained from (9.7.13) and (9.9.3) as

$$M_y/\rho_0 U^2 = \alpha\{S(1) - V\} + O(t^5 \log^2 t), \qquad (9.9.6)$$

where V is the volume of the body. This moment tends to decrease the incidence. The centre of lift is at a distance

$$1 - S_m/S(1) \qquad (9.9.7)$$

downstream from the nose, where S_m is the mean area of cross-section of the body.

These results apply to both subsonic and supersonic flows, and were first given by Tsien (1938) for supersonic flows.

The drag in subsonic flow vanishes in the only case for which it can be calculated on the present theory (i.e. when the body is pointed at both ends).

For supersonic flow, the potential ϕ_0 on the body, where

$$\zeta-\zeta_g=R(z)\,e^{i\theta_0}, \tag{9.9.8}$$

say, is the real part of

$$w_0=b_0+\frac{1}{2\pi}S'(z)\{\log R(z)+i\theta_0\}+\alpha R(z)\,e^{-i\theta_0}, \tag{9.9.9}$$

that is,

$$\phi_0=\frac{1}{2\pi}S'(z)\log\tfrac{1}{2}BR(z)-\frac{1}{2\pi}\int_0^z S''(s)\log(z-s)\,ds+\alpha R(z)\cos\theta_0. \tag{9.9.10}$$

From the boundary condition on the body,

$$\frac{\partial\phi_0}{\partial\nu}=R'(z)-\alpha\cos\theta_0, \tag{9.9.11}$$

and hence

$$\left(\int_C\phi_0\frac{\partial\phi_0}{\partial\nu}\,d\tau\right)_{z=1}$$
$$=\int_0^{2\pi}\left\{\left(\frac{S'(1)}{2\pi}\log\tfrac{1}{2}BR(z)-\frac{1}{2\pi}\int_0^1\log(1-s)\,S''(s)\,ds\right)R'(1)\right.$$
$$\left.-\alpha^2R(1)\cos^2\theta_0\right\}R(1)\,d\theta_0$$
$$=\frac{1}{2\pi}\{S'(1)\}^2\log\tfrac{1}{2}BR(1)-\frac{S'(1)}{2\pi}\int_0^1\log(1-s)\,S''(s)\,ds-S(1)\alpha^2. \tag{9.9.12}$$

On substituting this expression in the general formula (9.8.7), the drag is given by

$$\frac{D}{\tfrac{1}{2}\rho_0U^2}=\frac{1}{2\pi}\int_0^1\int_0^1\log\frac{1}{|z-s|}\,S''(s)\,S''(z)\,ds\,dz$$
$$-\frac{S'(1)}{\pi}\int_0^1\log\frac{1}{1-z}\,S''(z)\,dz+\frac{1}{2\pi}\{S'(1)\}^2\log\frac{2}{BR(1)}$$
$$+S(1)\,\alpha^2+O(t^6\log^2 t). \tag{9.9.13}$$

It is shown later, in §9.12, that this formula is a special case of a more general result which applies to a body of revolution whose meridian profile has discontinuities of slope.

9.10. Plane wings of small aspect-ratio

The condition on the maximum curvature of any section of a body prevents the study of some important problems, and it is desirable to consider the effect of removing this condition for certain special cases. Such a special case is that of a plane wing of small aspect-ratio, with highly swept-back leading edges, which is considered in this section.

The plane wing may be taken to be the limiting form of a slender body of elliptical cross-section as the eccentricity of the ellipses tends to unity. If the eccentricity is nearly unity, then for a finite incidence, the velocity in the vicinity of the leading edges becomes very large. However, for a fixed eccentricity, it is always possible to find an incidence such that the perturbation velocity is small enough to keep the approximations valid, and so for all smaller incidences. Thus quantities such as $(\partial C_L/\partial\alpha)_{\alpha=0}$ and $(\partial^2 C_D/\partial\alpha^2)_{\alpha=0}$ can be calculated to a known approximation, and by letting the eccentricity tend to unity, these quantities remain finite. On the other hand, if the order of the above two limiting processes is reversed, the velocity becomes infinite at the points where the curvature becomes infinite, for any finite value of the incidence. The solutions for the potential ϕ_0 and its derivatives are nevertheless a true approximation to the exact potential except in the immediate vicinities of the points of infinite curvature. In particular, the pressure is inaccurate only for comparatively small regions on the body surface, and it can be assumed that the overall aerodynamic forces are given correctly if calculated by the momentum-flux method. The discussion in §4.6 applies equally to the above case.

In what follows, the above assumption is used to calculate the body forces when there is no restriction on the curvature of sections, using the formulae obtained previously, but omitting the O-terms since these have lost some of their meaning.

Consider now a plane wing, pointed at its upstream end, with highly swept-back leading edges (in order to confirm with the condition for a slender body) and lying in the plane $x=-\alpha z$, so that its incidence is α. Let $a(z)$ be the semi-span at any section $z=constant$, and let the origin of ζ_1 ($\zeta_1=x_1+iy_1=\zeta+\alpha x$) be taken at the centre of the local span. Then the appropriate incompressible flow in two dimensions is that for a flat plate with its edges at $y_1 = \pm a(z)$, and the complex potential is

$$w_0 = \alpha\{\sqrt{(\zeta_1^2 + a^2)} - \zeta_1\}. \qquad (9.10.1)$$

The flow given from this expression for w_0 is appropriate so long as the edges of the wing are leading edges. When the edges are trailing edges, this form for w_0 is no longer correct, since a vortex sheet is formed downstream from the trailing edges. The strength of this vortex sheet is determined from the Kutta-Joukowski condition and from the condition of constant vorticity along each vortex line. (In general, these vortex lines can no longer be taken as straight, as in fully linearized theory, but must follow the streamlines.) Thus, in principle, other terms must be added to (9.10.1), representing the effect of the vortices, and such that the singularities in the velocity at $y_1 = \pm a(z)$ are cancelled. If the incidence is small, these vortices will lie approximately in the plane of the wing (actually they are found to do so exactly in this case) and so approximately on the imaginary axis in any ζ_1-plane; i.e. between $\zeta_1 = \pm ib$, when $2b$ is the maximum span of the wing. The potential w_0 given by

$$w_0 = \alpha\{\sqrt{(\zeta_1^2 + b^2)} - \zeta_1\} \qquad (9.10.2)$$

satisfies the boundary condition on any part of the wing, which must lie inside the limits $\zeta_1 = \pm ib$. Since b is a constant, the corresponding ϕ_0 is independent of z, for constant ζ_1; also the magnitude of the perturbation velocity is an even function of x_1, and hence there is no pressure difference across the vortex sheet (nor across the wing). Thus w_0 given by (9.10.2) satisfies all the boundary conditions, and is the same as the potential for the flow due to a flat plate of width $2b$.

On putting $\zeta_1 = \zeta + \alpha z$ in (9.10.1) and (9.10.2), and expanding in inverse powers of ζ, it is found that

$$a_1(z) = \tfrac{1}{2}\{a(z)\}^2 \quad (a'(z) \geqslant 0) \qquad (9.10.3)$$

for sections upstream from the maximum span, and

$$a_1(z) = \tfrac{1}{2}b^2 \quad (a(z) \leqslant b) \qquad (9.10.4)$$

for sections downstream from the maximum span.

The general results of §9.9 are still valid if C at $z = 1$ includes both sides of any vortex sheets. Thus, if b is the maximum semi-span (whether it occurs at $z=1$ with $da/dz \geqslant 0$, or for some $z<1$), the lift is obtained from (9.7.10) as

$$F_x = \pi \alpha \rho_0 U^2 b^2, \quad F_y = 0, \tag{9.10.5}$$

and

$$C_L = \frac{F_x}{\frac{1}{2}\rho_0 U^2 S_w} = \frac{1}{2}\pi A \alpha, \tag{9.10.6}$$

where S_w is the wing planform area, and A is the aspect-ratio ($A = 4b^2/S_w$). From (9.8.17) the drag coefficient is

$$C_D = \frac{D}{\frac{1}{2}\rho_0 U^2 S_w} = \frac{1}{4}\pi A \alpha^2. \tag{9.10.7}$$

It is to be noticed that there is no lift on sections downstream from the maximum span. This is due to the fact that the downwash caused by the vortex sheet is just the velocity required to satisfy the boundary condition on the wing. In fact, the downwash is $U\alpha$ over both wing and vortex sheet, so the vortex sheet remains coplanar with the wing everywhere, including the portion downstream from the wing. The shape of the wing planform downstream from the section of maximum span is thus shown to be immaterial, for the purposes of this theory, so the condition of slenderness need not be imposed on it.

It is interesting to compare (9.10.6) and (9.10.7) with the corresponding results of § 8.11 for a rectangular wing of small aspect-ratio. The lift coefficient for the rectangular wing is given correctly by slender-body theory, but the drag is not; the discrepancy arises from the fact that (9.10.7) includes the effect of suction on subsonic leading edges, and this suction is absent in the case of a rectangular wing.

The results of this section were obtained originally by R. T. Jones (1946*a*).

9.11. Winged bodies of revolution

The complex potential w_0 appropriate to the case of a body of revolution carrying plane wings of small aspect-ratio can be obtained from the results of the previous section by a simple conformal transformation.

Consider a slender pointed body of revolution of radius $R(z)$ at any section, at incidence α to the stream, so that $\zeta_g = -\alpha z$, and carrying wings symmetrically placed on either side of the body in the plane $x = -\alpha z$. Let the edges of the wings be at a distance $a(z)$ from the body axis at any section. In what follows the functional dependence of R and a upon z will be omitted.

At zero incidence, when the axis of the body is in the direction of the main stream, the flow is that for the body alone, since the wings lie in stream surfaces. At incidence α, terms due to the cross-flow over the body and wings have to be added to the complex potential for zero incidence. These extra terms can be obtained from expressions of the form (9.10.1) by replacing ζ_1 by $(\zeta_1 + R^2/\zeta_1)$ and a by $(a - R^2/a)$ in the square root; the origin of ζ_1 is again taken at $\zeta = -\alpha z$. Thus the complete complex potential for this case is

$$w_0 = b_0 + \frac{1}{2\pi} S'(z) \log \zeta_1 + \alpha\{\sqrt{[(\zeta_1 + R^2/\zeta_1)^2 + (a - R^2/a)^2]} - \zeta_1\}. \tag{9.11.1}$$

This form for w_0 is appropriate when the edges of the wings are leading edges. The definition of a leading edge in this case is not quite as simple as it is for the plane wing alone, since the streamlines are curved if R is not constant. From (9.11.1), it can be shown that $\partial\phi_0/\partial y_1$ in the plane of the wings, for $y_1 > a$, is

$$\left(\frac{\partial \phi_0}{\partial y}\right)_{x_1=0,\, y_1>a} = \frac{S'(z)}{2\pi y_1}; \tag{9.11.2}$$

this quantity is the slope of the projection of the streamlines near to the plane $x_1 = 0$ on to this plane, hence the equations of the streamlines near the plane of the wings are approximately

$$y_1^2 = R^2 + \text{constant}. \tag{9.11.3}$$

This result can be used to find the value of z for which the edge of the wing is tangent to a streamline. Upstream from this point the edge is a leading edge, and downstream from this point the edge is a trailing edge.

As for the case of the wing alone in the previous section, there is a vortex sheet extending downstream from the trailing edge if the Kutta-Joukowski condition is applied. The determination of the vorticity distribution in this sheet for a general variation of R with z is tedious, and in general the expression for w_0 contains hyper-elliptic integrals. However, when R is constant downwstream from the section of maximum span, the potential w_0 is easily obtained from (9.10.2); for if b is the distance of the wing edge from the centre of the body at maximum span, then the conformal transformation used in obtaining (9.11.1), with a replaced by b, gives

$$w_0 = \alpha\{\sqrt{[(\zeta_1 + R^2/\zeta_1)^2 + (b - R^2/b)^2]} - \zeta_1\}, \qquad (9.11.4)$$

since $S'(z)$ is now zero.

The conformal transformation by means of which (9.11.4) is obtained from (9.10.2) does not alter the geometry of the vortex system, which remains coplanar with the wings. Thus, if $R = constant$, the boundary condition on the wing is still satisfied by the downwash from the vortex sheet; again the shape of the wing planform is immaterial downstream from the section of maximum span, provided that $a \leqslant b$, and there is no lift on these sections. These results are true only if $R = constant$; if R varies, then, in general, there is a lift force downstream from the maximum span.

In general, the dimensions of the body and wings (i.e. R and a) must be continuous bounded functions of z together with their first derivatives, otherwise $\partial^2 w_0/\partial z^2$ becomes infinite, the pressure is discontinuous all over any section of discontinuity, and the analysis given here breaks down. This condition has already been shown to be unnecessary downstream from the maximum span, and it can be shown that, in addition, it is not necessary on the leading edges when $a = R$, so da/dz may have a discontinuity at this point. By differentiating w_0 given by (9.11.1) it can be shown that $\partial w_0/\partial z$ is continuous, and $\partial^2 w_0/\partial z^2$ remains finite when $a = R$ if da/dz is discontinuous there. Thus the leading edges of the wing can come from the body at a finite angle without the necessity for introducing further approximations.

The lift for winged bodies of revolution of two different types can now be determined easily: (i) a body of revolution having general variation of radius of cross-section with wings whose edges are everywhere leading edges upstream from the base (see fig. 9.2 *a*), and (ii) a body of revolution with a uniform section downstream from the section of maximum wing span (see fig. 9.2 *b*). For subsonic flow, the drag can only be determined in case (i) when the body is pointed at the rear end. For supersonic flow, the drag can be calculated in both cases.

If $2b$ is the maximum span, and R_1 is the radius of the base in each case, then the lift forces, and the extra drag forces due to incidence in supersonic flow, have the same form for both. From (9.11.1) and (9.11.4), by putting $\zeta_1=\zeta+\alpha z$ and expanding in inverse powers of ζ, the coefficient of $1/\zeta$ at $z=1$ is found to be

$$a_1(1)=\frac{b^4+R_1^4}{2b^2}-\frac{\zeta_0(1)\,S'(1)}{2\pi}\quad\text{from (9.11.1),}\qquad(9.11.5)$$

and

$$a_1(1)=\frac{b^4+R_1^4}{2b^2}\quad\text{from (9.11.4).}\qquad(9.11.6)$$

By substituting these values for $a_1(1)$ in (9.7.10), it is found that, in both cases, the lateral force is a lift force only, given by

$$F_x=\pi\alpha\rho_0U^2b^2\left(1-\frac{R_1^2}{b^2}+\frac{R_1^4}{b^4}\right),\quad F_y=0.\qquad(9.11.7)$$

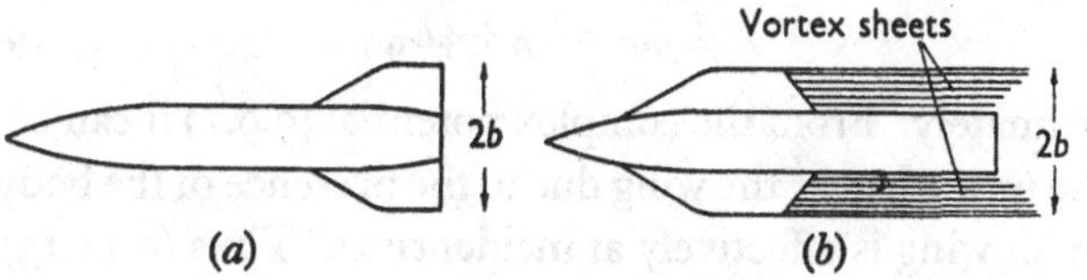

Fig. 9.2. Illustrations of the two classes of winged bodies of revolution for which the lift forces are given by (9.11.7).

The drag force in supersonic flow is then given from (9.8.16) by

$$D=D_0+\tfrac{1}{2}\pi\rho_0U^2b^2\left(1-\frac{R_1^2}{b^2}+\frac{R_1^4}{b^4}\right)\alpha^2,\qquad(9.11.8)$$

where D_0 is the drag at zero incidence. For subsonic flow, in case (i) with $R_1=0$ only, the drag force is

$$D=\tfrac{1}{2}\pi\rho_0U^2b^2\alpha^2,\qquad(9.11.9)$$

since $D_0=0$.

The lift given by (9.11.7) may be compared with that for a wing of small aspect-ratio given by (9.10.5). Since $R_1/b<1$, it is seen that the lift for the winged body is less than that for a plane wing of the same maximum span, showing that the body is less effective as a producer of lift than a wing of the same projected planform. It is interesting to notice that this statement is only true when the body carries wings; for the body alone is just as effective as the wing of the same projected planform, as is seen by putting $b=R_1$ in (9.11.7) for example.

In case (ii), when the wing root lies wholly on the cylindrical portion of the body, the difference between the lift, L_{BW}, on the combined system, and the lift, L_B, for the body alone is

$$L_{BW}-L_B=\pi\alpha\rho_0 U^2b^2(1-R_1^2/b^2)^2. \qquad (9.11.10)$$

Now the lift L_W on the wing alone (of maximum span $2(b-R_1)$) is, from (9.10.5),

$$L_W=\pi\alpha\rho_0 U^2(b-R_1)^2; \qquad (9.11.11)$$

hence, by combining these two results,

$$L_{BW}=L_B+(1+R_1/b)^2 L_W, \qquad (9.11.12)$$

which is a formula for estimating the lift of a wing-body combination in terms of the lifts of each component by itself.

For a very small wing, when $b-R_1$ is small, (9.11.12) gives

$$L_{BW}=L_B+4L_W, \qquad (9.11.13)$$

approximately. From the complex potential (9.9.1) it can be shown that the upwash near the wing due to the presence of the body is $U\alpha$, so a small wing is effectively at incidence 2α. Thus (9.11.13) shows that half of the extra lift comes from the wing and half from the body in this case.

9.12. Supersonic flow past bodies of revolution when $S'(z)$ is discontinuous

When the slope of the meridian section of a body of revolution is discontinuous, the slender-body approximation in the form given in the previous sections of this chapter ceases to be valid, and the formula (9.9.13) gives an infinite value for the drag force even when the discontinuities are smoothed out by small curved arcs whose curvatures are allowed to increase indefinitely. This difficulty in slender-body theory has been overcome by Lighthill (1948*b*), who has shown that the potential on and near the body surface just downstream from a discontinuity cannot be represented by a series of the form (9.3.15), and that additional terms are necessary. Since the streamlines on the body surface change direction discontinuously at each discontinuity, it is to be expected that there are consequent discontinuities in pressure quite analogous to those found in Chapter 8 for quasi-cylindrical ducts, and hence that a similar operational approach should give the required corrections.

In the following theory, the body under consideration may be ducted and have a sharp-edged open nose of finite radius, in which case it is assumed that the fluid which enters the body through the intake continues to move with supersonic speed for at least a short distance inside the body, so that the external and internal flows are independent. This assumption is justified if the internal flow is not choked to such an extent that 'spill-over' occurs at the intake. If 'spill-over' does occur, then the problem of determining the flow cannot be treated by linearized theory. With the z-axis coincident with the axis of the body, let $R(z)$ be the radius of the body at any section $z = constant$; let the body be of unit length and lie in the interval $0 \leqslant z \leqslant 1$, and let $R(z)$ be a continuous function of z in this interval such that $R(z) < At$ and $|R'(z)| < At$ (cf. (9.2.6)). Let $R'(z)$ possess a finite number of discontinuities at $z = z_i$ $(i = 0, 1, \ldots, n)$; let $R(z)$ be analytic in each interval $0 = z_0 < z < z_1$, $z_1 < z < z_2$, $\ldots$, $z_{n-1} < z < z_n = 1$, and let $R(z_i) = R_i$ and $R'(z_i + 0) - R'(z_i - 0) = \eta_i$, with the convention that $R'(z) = 0$ for $z < 0$ and $z > 1$. For the operational solution given below, it is convenient to rotate the co-ordinate frame with the body when applying incidence, so that the z-axis always coincides with the body axis.† The effect of small positive incidence, α, is then equivalent to a cross-stream with velocity $U\alpha$ in the direction of the x-axis, and from the results of § 1.12, the linearized equations are still valid for small disturbances on the combined main stream and cross-stream; alternatively the cross-stream may be considered to be part of the perturbation velocity, since it is irrotational.

The boundary condition on the body surface is

$$\left(\frac{\partial \phi_0}{\partial r}\right)_{r=R(z)} = R'(z) - \alpha \cos\theta, \qquad (9.12.1)$$

and the appropriate operational solution for the potential is

$$\phi_0 = \phi_{00} + \phi_{01},$$

where

$$\phi_{00} = a_0(p)\, K_0(Bpr) \quad \text{and} \quad \phi_{01} = a_1(p)\, K_1(Bpr) \cos\theta. \qquad (9.12.2)$$

† This is not necessary for a quasi-cylindrical duct, because then there is a mean body surface on which the boundary conditions can be applied; but slender bodies do not possess such mean surfaces, by definition (see §1.9).

Hence for *constant* r it follows as in Chapter 8, that

$$\frac{\partial\phi_{00}}{\partial z}=-\frac{K_0(Bpr)}{BK_1(Bpr)}\frac{\partial\phi_{00}}{\partial r}\quad\text{and}\quad\frac{\partial\phi_{01}}{\partial z}=\frac{K_1(Bpr)}{BK_1'(Bpr)}\frac{\partial\phi_{01}}{\partial r},\tag{9.12.3}$$

or, on interpretation by the product theorem,

$$\frac{\partial\phi_{00}(r,z)}{\partial z}=-\frac{1}{B}\int_{s=-0}^{s=z}V_0'\left(\frac{z-s}{Br}\right)\mathrm{d}\left(\frac{\partial\phi_{00}(r,s)}{\partial r}\right),\tag{9.12.4}$$

and

$$\frac{\partial\phi_{01}(r,z)}{\partial z}=-\frac{1}{B}\int_{s=-0}^{s=z}V_1'\left(\frac{z-s}{Br}\right)\mathrm{d}\left(\frac{\partial\phi_{01}(r,s)}{\partial r}\right),\tag{9.12.5}$$

where the integrals are put into Stieltjes form because the radial component of velocity is discontinuous, and the lower limit of integration represents any station upstream from the region in which the interpretations of the operational forms (9.12.3) do not vanish. From these formulae, the value of $\partial\phi_0/\partial z$ on the body (when $r=R(z)$) can be found if the values of $\partial\phi_{00}/\partial r$ and $\partial\phi_{01}/\partial r$ at $r=R(z)$ but at $z=s$, can be found from the known values at $r=R(s)$, $z=s$, given by the boundary condition (9.12.1). For values of z that are not at small distances $O(t)$ downstream from the discontinuities, the radial derivatives of ϕ_{00} and ϕ_{01} on and near the body surface are given approximately by expanding the operational forms for small p, in which case the boundary condition (9.12.1) yields the results

$$\frac{\partial\phi_{00}}{\partial r}=\frac{S'(z)}{2\pi r}+O(t^3\log t)\quad\text{and}\quad\frac{\partial\phi_{01}}{\partial r}=-\frac{S(z)}{\pi r^2}\alpha\cos\theta+O(t^3\log t),\tag{9.12.6}$$

which are the values given by ordinary slender-body theory (cf. §9.9). However, on and very near the body surface at distances $O(t)$ downstream from the discontinuities, these results are still correct within a factor $1+O(t)$ because the variation with r in this region is not very different from that given; in fact, the variation is like $1/\sqrt{r}$ for that part of the radial derivative which arises from each discontinuity (cf. (8.3.12)), and as given in (9.12.6) for the remainder. Thus, to a sufficient order of approximation,

$$\left(\frac{\partial\phi_{00}}{\partial z}\right)_{r=R(z)}=-\frac{1}{2\pi BR(z)}\int_{-0}^{z}V_0'\left(\frac{z-s}{BR(z)}\right)\mathrm{d}S'(s),\tag{9.12.7}$$

and

$$\left(\frac{\partial\phi_{01}}{\partial z}\right)_{r=R(z)}=\frac{\alpha\cos\theta}{\pi B\{R(z)\}^2}\int_{-0}^{z}V_1'\left(\frac{z-s}{BR(z)}\right)\mathrm{d}S(s).\tag{9.12.8}$$

On evaluating the terms arising from the discontinuities in $S'(z)$, and then integrating by parts, these formulae become

$$\left(\frac{\partial\phi_{00}}{\partial z}\right)_{r=R(z)} = -\frac{1}{2\pi}\int_{-0}^{z} V_0\left(\frac{z-s}{BR(z)}\right) dS''(s) - \sum_{i=0}^{n} \frac{R_i\eta_i}{BR(z)} V_0'\left(\frac{z-z_i}{BR(z)}\right), \tag{9.12.9}$$

and
$$\left(\frac{\partial\phi_{01}}{\partial z}\right)_{r=R(z)} = \frac{\alpha\cos\theta}{R(z)}\left\{\frac{1}{\pi}\int_0^z V_1\left(\frac{z-s}{BR(z)}\right) S''(s)\,ds + \frac{R_0^2}{BR(z)} V_1'\left(\frac{z}{BR(z)}\right) + 2\sum_{i=0}^{n} R_i\eta_i V_1\left(\frac{z-z_i}{BR(z)}\right)\right\}, \tag{9.12.10}$$

where the functions V_0, V_1, V_0', V_1' vanish for negative values of their arguments. By using the asymptotic values of the functions V_0, V_1, V_0', V_1', given in §8.5, it can be shown that, for the purpose of calculating aerodynamic forces, no further error in the order of approximation is made by writing these last two formulae as

$$\left(\frac{\partial\phi_{00}}{\partial z}\right)_{r=R(z)} = \frac{1}{2\pi} S''(z)\log\tfrac{1}{2}BR(z) - \frac{1}{2\pi}\int_{-0}^{z}\log(z-s)\,dS''(s) - \sum_{i=0}^{n}\frac{\eta_i}{B} V_0'\left(\frac{z-z_i}{BR_i}\right), \tag{9.12.11}$$

and
$$\left(\frac{\partial\phi_{01}}{\partial z}\right)_{r=R(z)} = \frac{\alpha\cos\theta}{R(z)}\left\{\frac{S'(z)}{\pi} + \frac{R_0}{B} V_1'\left(\frac{z}{BR_0}\right) - 2\sum_{i=0}^{n} R_i\eta_i\left[1 - V_1\left(\frac{z-z_i}{BR_i}\right)\right]\right\}, \tag{9.12.12}$$

where each term in square brackets vanishes when z is less than the corresponding z_i. These are equivalent to the expressions given originally by Lighthill (1948*b*), and have the advantage of being analytically simpler than the equivalent expressions from which they are derived; clearly they represent only one pair of many possible equivalent formulae.

The expressions (9.12.11) and (9.12.12) show how the results of slender-body theory are modified by the presence of discontinuities, η_i, in $R'(z)$, and by the finite nose radius R_0. Reference to §8.5 shows that $V_0'(0) = V_1'(0) = 1$ and $V_1(0) = 0$, that $V_0'(z)$ differs appreciably from $1/z$ only for $0 < z < 3$, say, and that $V_1'(z)$ and $1 - V_1(z)$ fall rapidly towards zero in much the same interval (in fact, they become slightly negative eventually). Thus the discontinuities cause sudden changes in $\partial\phi_0/\partial z$ on the body, of amount $(\eta_0 - \alpha\cos\theta)/B$ at the nose when this has a finite radius, and of amount η_i/B at $z=z_i$ ($i \neq 0$), in agreement with two-dimensional theory and the theory of quasi-cylindrical ducts; also the effect of each discontinuity is damped out in a short distance, $O(t)$, downstream, leaving the value of $\partial\phi_0/\partial z$ at what it would have been if the discontinuity had been smoothed out. Since the pressure discontinuities are $O(t)$, they can have an appreciable effect on the drag; this effect increases the drag for very small values of the thickness parameter, t, but may decrease it for larger values of t.

The external drag, lift and moment can be found by integrating the pressures over the body, and using the asymptotic values of V_0, V_1, V_0', V_1' to find approximations to the integrals (see Lighthill, 1948*b*). The quadratic approximation to Bernoulli's equation has to be used, which in this case is

$$\frac{p-p_0}{\frac{1}{2}\rho_0 U^2} = -2\frac{\partial\phi_0}{\partial z} - \left(\alpha\cos\theta + \frac{\partial\phi_0}{\partial r}\right)^2 - \left(-\alpha\sin\theta + \frac{1}{r}\frac{\partial\phi_0}{\partial\theta}\right)^2 + \alpha^2.\dagger \tag{9.12.13}$$

The final result for the external drag is (Lighthill, 1948*b*)

$$\begin{aligned}\frac{D}{\frac{1}{2}\rho_0 U^2} = {} & \frac{1}{2\pi}\int_0^1\int_0^1 \log\frac{1}{|z-s|}\, S''(s)\, S''(z)\, ds\, dz \\ & + 2\sum_{i=0}^{n} R_i\eta_i \int_0^1 \log\frac{1}{|z-z_i|}\, S''(z)\, dz \\ & + 4\pi \sum_{i>j=0}^{n}\sum^{n} R_i R_j \eta_i\eta_j \log\frac{1}{z_i - z_j} + 2\pi\sum_{i=0}^{n} R_i^2\eta_i^2 \log\frac{2}{BR_i} \\ & + \{S(0)+S(1)\}\alpha^2,\end{aligned} \tag{9.12.14}$$

and the external lift and moment are the same as in ordinary slender body theory, i.e. as given by (9.9.4) and (9.9.6).

The formula (9.12.7) was originally given in a slightly different form by Lighthill,‡ who, guided by the form of (9.12.9), wrote

$$\left(\frac{\partial\phi_{00}}{\partial z}\right)_{r=R(z)} = -\frac{1}{2\pi B}\int_{-0}^{z} V_0'\left(\frac{z-s}{BR(s)}\right)\frac{dS'(s)}{R(s)}. \tag{9.12.15}$$

That this result differs from (9.12.7) only by a factor $1+O(t)$ can be shown by using the asymptotic value of V_0' and the fact, noted already, that the body radius changes only by $O(t^2)$ in a distance $O(t)$ downstream from each discontinuity. Numerically it is probably slightly more accurate than (9.12.7).

† When calculating the drag, it is sufficient to use the slender body result $(\partial\phi_0/r\partial\theta)_{r=R(z)} = -\alpha\sin\theta$.

‡ Unpublished communication.

9.13. Ducted bodies of revolution with annular intakes

The supersonic flow past a ducted body of revolution with an annular intake following a pointed fore-body can be found from the results of the previous section. Just as in the case of a nose intake, it has to be assumed that the external and internal flows are independent, an assumption that is justified if there is no 'spill-over' and if the effect of the boundary layer on the fore-body is not such as to cause instability of the flow. As usual, the displacement effects of this boundary layer are neglected, so the theory is not directly applicable when these are appreciable, as they may be for narrow intakes.

As before, consider a body of unit length, lying in $0 \leqslant z \leqslant 1$, which has an annular intake at $z = z_1$, where the z-axis coincides with the axis of the body. Let $R(z)$ be the radius of the body, such that $R(z) < Atz$ and $|R'(z)| < At$, and let $R(z)$ be analytic in the intervals $0 < z < z_1$, $z_1 < z < 1$.† Also let the intake be specified by

$$R(z_1 - 0) = R_0, \quad R(z_1 + 0) = R_1 > R_0,$$
$$R'(z_1 - 0) = \eta_0 \quad \text{and} \quad R'(z_1 + 0) = \eta_1.$$

Then the discontinuities in $S(z)$ and $S'(z)$ at $z = z_1$ are respectively $\pi(R_1^2 - R_0^2)$ and $2\pi(R_1\eta_1 - R_0\eta_0)$, and (9.12.7) and (9.12.8) give

$$\left(\frac{\partial \phi_{00}}{\partial z}\right)_{r=R(z)} = \frac{1}{2\pi} S''(z) \log \tfrac{1}{2}BR(z) - \frac{1}{2\pi}\int_{-0}^{z} \log(z-s)\,\mathrm{d}S''(s)$$
$$- \frac{R_1\eta_1 - R_0\eta_0}{BR_1} V_0'\left(\frac{z - z_1}{BR_1}\right), \quad (9.13.1)$$

and
$$\left(\frac{\partial \phi_{01}}{\partial z}\right)_{r=R(z)} = \frac{\alpha\cos\theta}{R(z)}\left\{\frac{S'(z)}{\pi} + \frac{R_1^2 - R_0^2}{BR_1} V_1'\left(\frac{z - z_1}{BR_1}\right)\right.$$
$$\left. - 2(R_1\eta_1 - R_0\eta_0)\left[1 - V_1\left(\frac{z - z_1}{BR_1}\right)\right]\right\}, \quad (9.13.2)$$

where V_0', V_1' and $1 - V_1$ vanish for $z < z_1$, which are equivalent to results given by Ward (1950).

† If there are other discontinuities in $R'(z)$, then the results of the previous section may be combined with those below to give the complete solution.

By integrating the pressures obtained from (9.12.13) over the external body surface (for details, see Ward, 1950), it is found that the external drag force is given by

$$\frac{D}{\frac{1}{2}\rho_0 U^2} = \frac{1}{2\pi}\int_0^1\int_0^1 \log\frac{1}{|z-s|} S''(s) S''(z)\, ds\, dz$$
$$-\frac{S'(1)}{\pi}\int_0^1 \log\frac{1}{1-z} S''(z)\, dz + \frac{\{S'(1)\}^2}{2\pi}\log\frac{2}{BR(1)}$$
$$+2(R_1\eta_1 - R_0\eta_0)\int_0^1 \log\frac{1}{|z-z_1|} S''(z)\, dz$$
$$-2(R_1\eta_1 - R_0\eta_0) S'(1)\log\frac{1}{1-z_1}$$
$$+2\pi(R_1\eta_1 - R_0\eta_0)^2\log\frac{2}{BR_1} + 2\pi R_0^2\eta_0^2\log\frac{R_1}{R_0}$$
$$+\{\pi(R_1^2 - R_0^2) + S(1)\}\alpha^2, \tag{9.13.3}$$

and that the external lift and moment are again the same as in ordinary slender-body theory.

From the linearized equations for subsonic flow, it can be verified that any asymmetrical disturbances at the ends of a cylindrical tube die away exponentially on proceeding into the tube. Hence, if the flow inside a ducted body of revolution is mainly subsonic, as it is usually in practice, and if no asymmetrical disturbance occurs near the exit at $z=1$, all the cross-momentum entering the duct is destroyed inside, which gives a lift force equal to

$$\rho_0\int_{R_0}^{R_1}\int_0^{2\pi}\left(\frac{\partial\Phi}{\partial x}\frac{\partial\Phi}{\partial z}\right)_{z=z_1-0} r\, dr\, d\theta. \tag{9.13.4}$$

The velocity components given by slender-body theory are valid at $z=z_1-0$, and by using these, (9.13.4) can be shown to equal

$$\rho_0 U^2\alpha\pi(R_1^2 - R_0^2). \tag{9.13.5}$$

The line of action of this lift force cannot be predicted in general without a detailed knowledge of the internal flow, which depends critically on the internal conditions, including combustion in the case of a propulsive duct, and its determination is outside the scope of this monograph. However, in some practical cases the internal flow becomes subsonic through a shock system almost immediately after entering the duct, and in such cases the line of action of the internal lift force may be expected to be at a distance $O(t)$ downstream from $z = z_1$. The internal pitching moment about the nose is then roughly

$$\rho_0 U^2 a \pi z_1 (R_1^2 - R_0^2), \tag{9.13.6}$$

in a sense tending to decrease the incidence.

Clearly the determination of the internal drag (or thrust) is well outside the scope of this monograph.

APPENDIX I

INTEGRAL IDENTITIES

Let $\mathbf{v}$ and $\mathbf{w}$ be two vector functions connected by the relation $\mathbf{w}=\boldsymbol{\Psi}.\mathbf{v}$, where $\boldsymbol{\Psi}$ is a constant symmetrical tensor, which are regular inside and on a closed surface, S, bounding a volume, V. Then the divergence theorem for the closed surface, S, applied to the tensor

$$\mathbf{vw}-\tfrac{1}{2}\mathbf{v}.\mathbf{w}\mathbf{I}, \tag{A. 1.1}$$

gives the quadratic integral identity

$$\int_S (\mathbf{vw}.\mathbf{n}-\tfrac{1}{2}\mathbf{v}.\mathbf{wn})\,dS=\int_V \{\mathbf{v}\nabla.\mathbf{w}+\mathbf{w}.\nabla\mathbf{v}-\tfrac{1}{2}\nabla(\mathbf{v}.\mathbf{w})\}\,dV$$

$$=\int_V \{\mathbf{v}\nabla.\mathbf{w}-\mathbf{w}\wedge(\nabla\wedge\mathbf{v})\}\,dV, \tag{A. 1.2}$$

where $\mathbf{n}$ is the unit outward normal to S.

The corresponding bi-linear identity, obtained by putting $\mathbf{v}=\mathbf{v}_1+\mathbf{v}_2$, $\mathbf{w}=\mathbf{w}_1+\mathbf{w}_2$, and $\mathbf{v}=\mathbf{v}_1-\mathbf{v}_2$, $\mathbf{w}=\mathbf{w}_1-\mathbf{w}_2$, where $\mathbf{w}_1=\boldsymbol{\Psi}.\mathbf{v}_1$, $\mathbf{w}_2=\boldsymbol{\Psi}.\mathbf{v}_2$, successively in (A. 1.2), and subtracting the two resulting identities, is

$$\int_S (\mathbf{v}_1\mathbf{w}_2.\mathbf{n}+\mathbf{v}_2\mathbf{w}_1.\mathbf{n}-\mathbf{v}_1.\mathbf{w}_2\mathbf{n})\,dS$$

$$=\int_V \{\mathbf{v}_1\nabla.\mathbf{w}_2+\mathbf{v}_2\nabla.\mathbf{w}_1-\mathbf{w}_1\wedge(\nabla\wedge\mathbf{v}_2)-\mathbf{w}_2\wedge(\nabla\wedge\mathbf{v}_1)\}\,dV, \tag{A. 1.3}$$

since $\mathbf{v}_1.\mathbf{w}_2=\mathbf{v}_2.\mathbf{w}_1$.

The identity (A. 1.2) is still true when the normal component of $\mathbf{v}$ and the tangential component of $\mathbf{w}$ are discontinuous on some surface, S' say, inside S.† For the integrand of the surface integral in (A. 1.2) can be written

$$\tfrac{1}{2}\mathbf{vw}.\mathbf{n}+\tfrac{1}{2}\mathbf{w}\wedge(\mathbf{v}\wedge\mathbf{n}), \tag{A. 1.4}$$

and if $\mathbf{v}'$ and $\mathbf{w}'$ are the discontinuities in $\mathbf{v}$ and $\mathbf{w}$ at S', then $\mathbf{n}\wedge\mathbf{v}'=0$ and $\mathbf{n}.\mathbf{w}'=0$, and the discontinuity in (A. 1.4) at S' is

$$\tfrac{1}{2}\mathbf{v}'\mathbf{w}.\mathbf{n}+\tfrac{1}{2}\mathbf{w}'\wedge(\mathbf{v}\wedge\mathbf{n})=\tfrac{1}{2}\mathbf{v}'\mathbf{w}.\mathbf{n}-\tfrac{1}{2}\mathbf{w}'.\mathbf{vn}=0, \tag{A. 1.5}$$

† The surface S' cannot be arbitrarily chosen, since if $\mathbf{n}$ is the unit normal to S' at any point, it can easily be shown that $\mathbf{n}$ must satisfy the condition $\mathbf{n}.\boldsymbol{\Psi}.\mathbf{n}=0$.

since $\mathbf{v}'$ is parallel to $\mathbf{n}$, and $\mathbf{\Psi}$ is symmetrical. By excluding S' from the interior of S by other surfaces that lie close to S' on both sides, applying (A. 1.2) to the new closed surfaces so obtained, and adding the results, it follows from (A. 1.5) that (A. 1.2) remains true provided that the volume integral is evaluated in the Lebesgue sense, which ignores the points of S' at which the derivatives of $\mathbf{v}$ and $\mathbf{w}$ are not defined.

The validity of (A. 1.3) under similar circumstances follows from its method of derivation from (A. 1.2).

APPENDIX 2

THE LINEARIZED EQUATIONS FOR NEARLY PARALLEL STEADY FLOWS

As an example of linearized equations for flows of a slightly more general character than those considered in this monograph, the linearized equations of motion for small perturbations of a parallel stream are developed below; such flows are sometimes called shear flows. If rectangular cartesian co-ordinates, x, y, z, are chosen so that the z-axis is parallel to the direction of the undisturbed stream, then this stream can be specified by the vector

$$\mathbf{U}(x,y) = U(x,y)\,\mathbf{k} \tag{A.2.1}$$

as a function of x and y only. If ρ_0 and p_0 are respectively the density and pressure of the undisturbed stream, then the equation of continuity for the undisturbed stream,

$$\mathbf{U}.\nabla\rho_0 + \rho_0\nabla.\mathbf{U} = 0, \tag{A.2.2}$$

shows that ρ_0 is constant on streamlines, and hence is a function of x and y only. The momentum equation for the undisturbed stream is

$$\rho_0\mathbf{U}.\nabla\mathbf{U} + \nabla p_0 = 0, \tag{A.2.3}$$

from which it follows that p_0 is constant over the whole flow field.

Now let $\mathbf{v}$ be a small perturbation of the parallel stream, with components u, v, w, so that the particle velocity is

$$\mathbf{u} = \mathbf{U} + \mathbf{v}, \tag{A.2.4}$$

and let ρ and p be the density and pressure respectively in the disturbed stream. If $\mathbf{v}$, $\rho-\rho_0$ and $p-p_0$, and their derivatives, are so small that their squares, products and higher powers can be neglected in the equations of motion, then the linearized equation of continuity is

$$\rho_0\nabla.\mathbf{v} + \mathbf{v}.\nabla\rho_0 + \mathbf{U}.\nabla(\rho-\rho_0) = \rho_0 Q, \tag{A.2.5}$$

the linearized momentum equation is

$$\rho_0\mathbf{U}.\nabla\mathbf{v} + \rho_0\mathbf{v}.\nabla\mathbf{U} + \nabla(p-p_0) = 0, \tag{A.2.6}$$

and the linearized condition of constant entropy along streamlines is

$$\mathbf{U}.\nabla(p-p_0) = c_0^2\{\mathbf{U}.\nabla(\rho-\rho_0) + \mathbf{v}.\nabla\rho_0\}, \tag{A.2.7}$$

where c_0 is the local speed of sound in the undisturbed stream, and is a function of x and y only.

The momentum equation, (A. 2.6), can be written in the form

$$\left(\mathbf{I}\frac{\partial}{\partial z}+\mathbf{\Phi}\right).\rho_0 U\mathbf{v} = -\nabla(p-p_0), \qquad \text{(A. 2.8)}$$

where $\mathbf{I}$ is the idemtensor, and the tensor $\mathbf{\Phi}$ is a function of x and y only, given by

$$U\mathbf{\Phi} = \frac{\partial U}{\partial x}\mathbf{ki} + \frac{\partial U}{\partial y}\mathbf{kj}, \qquad \text{(A. 2.9)}$$

from which is it easily seen that $\mathbf{\Phi}^2 = 0$. The integrating factor for (A. 2.8) is $\mathbf{I}+z\mathbf{\Phi}$, and, on multiplication by this factor and integration from $z=-\infty$ to z, (A. 2.8) gives

$$\rho_0 U\mathbf{v}(x,y,z) = \rho_0 U\mathbf{v}(x,y,-\infty) - (\mathbf{I}-z\mathbf{\Phi}).\int_{-\infty}^{z}(\mathbf{I}+z'\mathbf{\Phi}).[\nabla(p-p_0)]_{z=z'}\,\mathrm{d}z'. \qquad \text{(A. 2.10)}$$

If $\mathbf{v}_1$ is written for $\mathbf{v}(x,y,-\infty)$, and ϕ is a scalar function given by

$$\phi = -\int_{-\infty}^{z}(p-p_0)\,\mathrm{d}z, \qquad \text{(A. 2.11)}$$

then (A. 2.10) becomes

$$\rho_0 U\mathbf{v} = \rho_0 U\mathbf{v}_1 + \nabla\phi - \int_{-\infty}^{z}(z-z')[\mathbf{\Phi}.\nabla(p-p_0)]_{z=z'}\,\mathrm{d}z'. \qquad \text{(A. 2.12)}$$

The form of this last equation suggests that the parallel stream $\mathbf{U}$ be replaced by the nearly parallel stream specified by

$$\mathbf{u}_1 = \mathbf{U}+\mathbf{v}_1, \qquad \text{(A. 2.13)}$$

in which case the new perturbation velocity, $\mathbf{v}'$, with components u', v', w', is given by

$$\rho_0 U\mathbf{v}' = \nabla\phi - \int_{-\infty}^{z}(z-z')[\mathbf{\Phi}.\nabla(p-p_0)]_{z=z'}\,\mathrm{d}z'. \qquad \text{(A. 2.14)}$$

The last term in this equation is parallel to the z-axis, so it follows from (A. 2.11) and (A. 2.14) that

$$[\rho_0 Uu', \rho_0 Uv', (p_0-p)] = \nabla\phi. \qquad \text{(A. 2.15)}$$

The equation satisfied by ϕ can be obtained by eliminating w and $(\rho-\rho_0)$ from (A. 2.5), the z-component of (A. 2.6), and (A. 2.7), which gives

$$\frac{\partial}{\partial x}\left(\frac{u}{U}\right)+\frac{\partial}{\partial y}\left(\frac{v}{U}\right)+(M_0^2-1)\frac{\partial}{\partial z}\left(\frac{p-p_0}{\rho_0 U^2}\right)=\frac{Q}{U}, \qquad \text{(A.2.16)}$$

where M_0 is the Mach number of the undisturbed parallel stream, and is a function of x and y only. For $\mathbf{v}_1$ to have any meaning, it must represent a possible steady perturbation of the parallel stream; then u and v in (A. 2.16) can be replaced by u' and v', and it follows from (A. 2.15) that ϕ satisfies

$$\frac{\partial^2\phi}{\partial x^2}+\frac{\partial^2\phi}{\partial y^2}+(1-M_0^2)\frac{\partial^2\phi}{\partial z^2}$$
$$+\frac{\partial}{\partial x}\left(\log\frac{p_0}{\rho_0 U^2}\right)\frac{\partial\phi}{\partial x}+\frac{\partial}{\partial y}\left(\log\frac{p_0}{\rho_0 U^2}\right)\frac{\partial\phi}{\partial y}=UQ, \quad \text{(A. 2.17)}$$

since p_0 is a constant. The quantity $p_0/\rho_0 U^2$ is a function of M_0 only, and in the case of a perfect gas with constant specific heats,

$$\gamma p_0/\rho_0 U^2 = M_0^{-2},$$

whence (A. 2.17) becomes

$$\frac{\partial^2\phi}{\partial x^2}+\frac{\partial^2\phi}{\partial y^2}+(1-M_0^2)\frac{\partial^2\phi}{\partial z^2}-\frac{2}{M_0}\frac{\partial M_0}{\partial x}\frac{\partial\phi}{\partial x}-\frac{2}{M_0}\frac{\partial M_0}{\partial y}\frac{\partial\phi}{\partial y}=UQ. \quad \text{(A. 2.18)}$$

The linearized boundary condition for ϕ at the surfaces of solid bodies can be written down at once, and involves only the x- and y-derivatives; therefore, if (A. 2.17) or (A. 2.18) can be solved, then the pressure can be found from the relation

$$p-p_0=-\partial\phi/\partial z, \quad \text{(A. 2.19)}$$

which is the linearized Bernoulli equation for shear flows. For flow past slender bodies it is necessary to use a quadratic approximation to Bernoulli's equation, which can be obtained from the exact equation (1.5.4) by expanding in series as for ordinary linearized theory.

It is interesting to notice that the equations (A. 2.17) and (A. 2.18) for ϕ involve the stream variables U, ρ_0 and p_0 through the Mach number, M_0, only; also, if $\mathbf{v}_1=0$, the boundary conditions can be shown to depend on M_0 and p_0 only, p_0 occurring only as a factor of proportionality in ϕ. Thus it appears that the solution, and the pressures, forces, etc., depend only on the original distribution of $\rho_0 U^2$, and not on the way ρ_0 and U are distributed individually.

Some problems in the linearized theory of shear flows have been considered by Chang and Chu, Lighthill, and Robinson;† these authors have not used the function ϕ as above, but have worked in terms of the pressure, p, which satisfies the same equation as ϕ when $Q=0$.

† Chang, C. C. and Chu, B. T. (1951), 'Linearized theory of subsonic, transonic and supersonic flow with assigned velocity gradient' (unpublished Report, Department of Aeronautics, The Johns Hopkins University).

Lighthill, M. J. (1950), 'Reflection at a laminar boundary layer of a weak steady disturbance to a supersonic stream, neglecting viscosity and heat conduction', *Quart. J. Mech. Appl. Math.* 3, 303–25.

Robinson, A. (1952), 'Non-uniform supersonic flow', *Quart. Appl. Math.* 10, 307–19.

BIBLIOGRAPHY

ACKERET, J. (1925). Über Luftkräfte auf Flügel, die mit grösserer als Schallgeschwindigkeit bewegt werden. *Z. Flugtech.* **16**, 72–4. (Transl. *Tech. Memor. Nat. Adv. Comm. Aero., Wash.*, no. 317.)

ACKERET, J. (1928). Über Luftkräfte bei sehr grossen Geschwindigkeiten inbesondere bei ebenen Strömungen. *Helv. phys. acta*, **1**, 301–22.

ADAMS, G. J. (1951). Theoretical damping in roll and rolling effectiveness of slender cruciform wings. *Tech. Notes Nat. Adv. Comm. Aero., Wash.*, no. 2270.

ADAMS, G. J. and DUGAN, D. W. (1952). Theoretical damping in roll and rolling moment due to differential wing incidence for slender cruciform wings and wing-body combinations. *N.A.C.A. Rep.* no. 1088.

ADAMS, M. C. (1951). Determination of shapes of boat tail bodies of revolution for minimum wave drag. *Tech. Notes Nat. Adv. Comm. Aero., Wash.*, no. 2250.

ADAMS, M. C. and SEARS, W. R. (1952). On an extension of slender-wing theory. *J. Aero. Sci.* **19**, 424–5.

ADAMS, M. C. and SEARS, W. R. (1953). Slender body theory—review and extension. *J. Aero. Sci.* **20**, 85–98.

ALLEN, H. J. and PERKINS, E. W. (1951). A study of effects of viscosity on slender inclined bodies of revolution. *N.A.C.A. Rep.* no. 1048.

BARTELS, R. C. F. and LAPORTE, O. (1948*a*). An investigation of the exact solutions of the linearized equations for the flow past conical bodies. *Bumblebee Rep.* no. 75.

BARTELS, R. C. F. and LAPORTE, O. (1948*b*). Supersonic flow past a delta wing at angles of attack and yaw. *Rep. Engng. Res. Inst., Ann Arbor*, no. CM-471.

BEANE, B. (1951). The characteristics of supersonic wings having biconvex sections. *J. Aero. Sci.* **18**, 7–20.

BRODERICK, J. B. (1949). Supersonic flow round pointed bodies of revolution. *Quart. J. Mech. Appl. Math.* **2**, 98–120.

BROWN, C. E. (1946). Theoretical lift and drag of thin triangular wings at supersonic speeds. *N.A.C.A. Rep.* no. 839.

BROWN, C. E. (1950). The reversibility theorem for thin airfoils in subsonic and supersonic flow. *N.A.C.A. Rep.* no. 986.

BROWN, C. E. and ADAMS, M. C. (1948). Damping in pitch and roll of triangular wings at supersonic speeds. *N.A.C.A. Rep.* no. 892.

BROWN, C. E. and PARKER, H. M. (1945). A method for the calculation of external lift, moment and pressure drag of slender open-nose bodies of revolution at supersonic speeds. *N.A.C.A. Rep.* no. 808.

BROWNE, S. H., FRIEDMAN, L. and HODES, I. (1948). A wing-body problem in a supersonic conical flow. *J. Aero. Sci.* **15**, 443–52.

BRYSON, A. E. (1953). The stability derivatives for a slender missile with application to a wing-body-vertical tail configuration. *J. Aero. Sci.* **20**, 297–308.

BURNS, J. C. (1951). Airscrews at supersonic forward speeds. *Aero. Quart.* **3**, 23–50.

BUSEMANN, A. (1935). Aerodynamischer Auftrieb bei Überschallgeschwindigkeit. *Atti di Convegni* 5, *Accademia d'Italia* (*Proc. Vth Volta Congress*), pp. 328–60; *Luftfahrtforschung*, **12**, 210–19.

BUSEMANN, A. (1943). Infinitesimale kegelige Überschallstromung. *Schr. Dtschen. Akad. Luftfahrtforschung*, 7B, 105–22. (Transl. *Tech. Memor. Nat. Adv. Comm. Aero., Wash.*, no. 1100.)

CHANG, C.-C. (1951). Applications of von Karman's integral method in supersonic wing theory. *Tech. Notes Nat. Adv. Comm. Aero., Wash.*, no. 2317.

CHAPMAN, D. R. (1952). Airfoil profiles for minimum pressure drag at supersonic velocities—general analysis with application to linearized supersonic flow. *N.A.C.A. Rep.* no. 1063.

CHESTER, W. (1953). Supersonic flow past wing-body combinations. *Aero. Quart.* **4**, 287–314.

COHEN, C. B. and EVVARD, J. C. (1948). Graphical method of obtaining theoretical lift distributions on thin wings at supersonic speeds. *Tech. Notes Nat. Adv. Comm. Aero., Wash.*, no. 1676.

COHEN, D. (1951). Formulas for the supersonic loading, lift and drag of flat swept-back wings with leading edges behind the Mach lines. *N.A.C.A. Rep.* no. 1050.

COLEMAN, T. F. (1948). Supersonic lift solutions obtained by extending the simple linearized conical flow theory. *Rep. N. Amer. Aviation Inc.* no. CM-440.

CRAMER, R. H. (1951). Interference between wing and body at supersonic speeds—theoretical and experimental determination of pressures on the body. *J. Aero. Sci.* **18**, 629–32.

DES CLERS, B. and CHANG, C.-C. (1951). On some special problems in linearized axially symmetric flow. *J. Aero. Sci.* **18**, 127–38.

EVVARD, J. C. (1950). Use of source distributions for evaluating theoretical aerodynamics of thin finite wings at supersonic speeds. *N.A.C.A. Rep.* no. 951. (Supersedes *Tech. Notes Nat. Adv. Comm. Aero., Wash.*, no. 1382 (1947) and others.)

FERRARI, C. (1937). Campi di corrente ipersonora attorno a solidi di rivoluzione. *Aerotechnica*, **17**, 507–18.

FERRARI, C. (1948). Interference between wing and body at supersonic speeds—theory and application. *J. Aero. Sci.* **15**, 317–36. (See also Bolz, R. E. (1950), *J. Aero. Sci.* **17**, 453.)

FERRARI, C. (1949). Interference between wing and body at supersonic speeds—note on wind-tunnel results and addendum to calculations. *J. Aero. Sci.* **16**, 542–6.

Flax, A. H. (1949). Relations between the characteristics of a wing and its reverse in supersonic flow. *J. Aero. Sci.* **16**, 496–504.

Flax, A. H. (1952*a*). General reverse flow and variational theorems in lifting surface theory. *J. Aero. Sci.* **19**, 361–74.

Flax, A. H. (1952*b*). Integral relations in the linearized theory of wing-body interference. *Rep. Cornell Aero. Lab.* no. CAL-45.

Fraenkel, L. E. (1951). The theoretical wave drag of some bodies of revolution. *Rep. Memor. Aero. Res. Coun., Lond.*, no. 2842.

Fraenkel, L. E. (1952). Supersonic flow past slender bodies of elliptic cross-section. *R.A.E. Tech. Note Aero*, no. 2466. (To be published in *Rep. Memor. Aero. Res. Coun., Lond.*)

Friedlander, F. G. (1951). On the half-plane diffraction problem. *Quart. J. Mech. Appl. Math.* **4**, 344–57.

Germain, P. (1949*a*). La théorie générale des mouvements coniques et ses applications à l'aérodynamique supersonique. *O.N.E.R.A. Rap.* no. 34.

Germain, P. (1949*b*). La théorie des mouvements homogènes et son application au calcul de certaines ailes delta en régime supersonique. *Rech. aéro.* **7**, 3–16.

Germain, P. and Bader, R. (1948). Théorie générale d'écoulement supersonique autour d'un obstacle aplati sur un plan. *O.N.E.R.A. Rap.* no. 1/1155A.

Glauert, H. (1928). The effect of compressibility on the lift of an aerofoil. *Proc. Roy. Soc.* A, **118**, 113–9.

Goldstein, S. and Ward, G. N. (1950). The linearized theory of conical fields in supersonic flow, with applications to plane aerofoils. *Aero. Quart.* **2**, 39–84.

Goldstein, S. and Young, A. D. (1943). The linear perturbation theory of compressible flow with application to wind tunnel interference. *Rep. Memor. Aero. Res. Coun., Lond.*, no. 1909.

Goldsworthy, F. A. (1952). Supersonic flow over thin symmetrical wings with given surface pressure distribution. *Aero. Quart.* **3**, 263–79.

Goodman, T. R. (1949). The lift distribution on conical and non-conical flow regions of thin finite wings in a supersonic stream. *J. Aero. Sci.* **16**, 365–74. (See also *ibid.* **16**, 703–4; **17**, 376–8.)

Göthert, B. (1940). Ebene und raumliche Strömungs bei hohen Unterschallgeschwindigkeit. *Lilienthal Ges. Luftfahrtforschung Ber.* **127**. (Transl. *Tech. Memor. Nat. Adv. Comm. Aero., Wash.*, no. 1105.)

Graham, E. W. (1952). A drag reduction method for wings of fixed planform. *J. Aero. Sci.* **19**, 823–5.

Gunn, J. C. (1947). Linearized supersonic aerofoil theory. *Phil. Trans.* A, **240**, 327–73.

Hadamard, J. (1923). *Lectures on Cauchy's Problem in Linear Partial Differential Equations.* New Haven: Yale University Press.

Harmon, S. M. (1949*a*). Stability derivatives at supersonic speeds of thin rectangular wings with diagonals ahead of tip Mach lines. *N.A.C.A. Rep.* no. 925.

Harmon, S. M. (1949*b*). Theoretical relations between the stability derivatives of a wing in direct and reverse flow. *Tech. Notes Nat. Adv. Comm. Aero., Wash.*, no. 1943.

Harmon, S. M. and Jeffreys, I. (1950). Theoretical lift and damping in roll of thin wings with arbitrary sweep and taper at supersonic speeds. Supersonic leading and trailing edges. *Tech. Notes Nat. Adv. Comm. Aero., Wash.*, no. 2114.

Hayes, W. D. (1946*a*). Linearized conical supersonic flow. *VIth Int. Congr. Appl. Mech.* (Paris) (unpublished). (See *Rep. N. Amer. Aviation Inc.* no. NA-46-818.)

Hayes, W. D. (1946*b*). Linearized supersonic flows with axial symmetry. *Quart. Appl. Math.* **4**, 255–61.

Hayes, W. D. (1947). Linearized supersonic flow. Thesis, Cal. Inst. Tech. *Rep. N. Amer. Aviation Inc.* no. AL-222.

Hayes, W. D. (1948). Reversed flow theorems in supersonic aerodynamics. *Proc. VIIth Int. Congr. Appl. Mech.* (London), **2**. (*Rep. N. Amer. Aviation Inc.* no. AL-755.)

Hayes, W. D., Browne, S. H. and Lew, R. J. (1947). Linearized theory of conical supersonic flow with applications to triangular wings. *Rep. N. Amer. Aviation Inc.* no. NA-46-818.

Hayes, W. D. and Linstone, H. A. (1948). A development of Evvard's supersonic wing theory. *Rep. N. Amer. Aviation Inc.* no. AL-746.

Hayes, W. D., Roberts, R. C. and Haaser, N. (1952). Generalized linearized conical flow. *Tech. Notes Nat. Adv. Comm. Aero., Wash.*, no. 2667.

Heaslet, M. A. and Lomax, H. (1947). The calculation of downwash behind supersonic wings with an application to triangular planforms. *Tech. Notes Nat. Adv. Comm. Aero., Wash.*, no. 1620.

Heaslet, M. A. and Lomax, H. (1948). The use of source-sink and doublet distributions extended to the solution of arbitrary boundary value problems in supersonic flow. *N.A.C.A. Rep.* no. 900.

Heaslet, M. A. and Lomax, H. (1950). The application of Green's theorem to the solution of boundary-value problems in linearized supersonic wing theory. *N.A.C.A. Rep.* no. 961.

Heaslet, M. A., Lomax, H. and Jones, A. L. (1947). Volterra's solution of the wave equation as applied to three-dimensional supersonic airfoil problems. *N.A.C.A. Rep.* no. 889.

Heaslet, M. A. and Spreiter, J. R. (1953). Reciprocity relations in aerodynamics. *N.A.C.A. Rep.* no. 1119.

Holt, M. (1950). The flow of two adjacent plane supersonic jets past flat-plate wings. *Quart. J. Mech. Appl. Math.* **3**, 200–16.

Holt, M. (1951). The flow of two adjacent plane supersonic jets past flat-plate wings. II. *Quart. J. Mech. Appl. Math.* **4**, 419–31.

JACK, J. R. (1950). Theoretical wave drags and pressure distributions for axially symmetric open-nosed bodies. *Tech. Notes Nat. Adv. Comm. Aero., Wash.*, no. 2115.

JONES, A. L. (1950). The theoretical lateral-stability derivatives for wings at supersonic speeds. *J. Aero. Sci.* **17**, 39–46.

JONES, A. L. and ALKSNE, A. (1950). A summary of lateral stability derivatives calculated for wing plan forms in supersonic flow. *N.A.C.A. Rep.* no. 1052.

JONES, R. T. (1946*a*). Properties of low-aspect-ratio pointed wings at speeds below and above the speed of sound. *N.A.C.A. Rep.* no. 835.

JONES, R. T. (1946*b*). Thin oblique airfoils at supersonic speeds. *N.A.C.A. Rep.* no. 851.

JONES, R. T. (1948). Subsonic flow over thin oblique airfoils at zero lift. *N.A.C.A. Rep.* no. 902.

JONES, R. T. (1950). Leading-edge singularities in thin airfoil theory. *J. Aero. Sci.* **17**, 307–10.

JONES, R. T. (1951). The minimum drag of thin wings in frictionless flow. *J. Aero. Sci.* **18**, 75–81.

JONES, R. T. (1952). Theoretical determination of the minimum drag of airfoils at supersonic speeds. *J. Aero. Sci.* **19**, 813–22.

JONES, R. T. and MARGOLIS, K. (1946). Flow over a slender body of revolution at supersonic velocities. *Tech. Notes Nat. Adv. Comm. Aero., Wash.*, no. 1081.

KARMAN, T. VON (1935). The problem of resistance in compressible fluids. *Atti di Convegni* 5, *Accademia d'Italia* (*Proc. Vth Volta Congress*), pp. 232–77.

KARMAN, T. VON (1947). Supersonic aerodynamics—principles and applications. *J. Aero. Sci.* **14**, 373–409.

KARMAN, T. VON and MOORE, N. B. (1932). Resistance of slender bodies moving with supersonic velocities. *Trans. Amer. Soc. Mech. Engrs.* **54**, 303–10.

KIRKBY, S. and ROBINSON, A. (1952). Interference on a wing due to a body at supersonic speeds. *Rep. Memor. Aero. Res. Coun., Lond.*, no. 2500.

KOLODNER, I. (1950). On the linearized theory of supersonic flows through axially symmetrical ducts. *Commun. Pure Appl. Math.* **3**, 133–52.

KUSSNER, H. G. (1940). Allgemeine Tragflächentheorie. *Luftfahrt-forschung*, **17**, 370–8. (Transl. *Tech. Memor. Nat. Adv. Comm. Aero., Wash.*, no. 979.)

LAGERSTROM, P. A. (1948). Linearized supersonic theory of conical wings. *Tech. Notes Nat. Adv. Comm. Aero., Wash.*, no. 1685.

LAGERSTROM, P. A. and GRAHAM, M. E. (1947). Downwash and sidewash induced by three-dimensional lifting wings in supersonic flow. *Rep. Douglas Aircraft Co.* no. SM-13007.

LAGERSTROM, P. A. and GRAHAM, M. E. (1949). Linearized theory of supersonic control surfaces. *J. Aero. Sci.* **16**, 31–4.

LAGERSTROM, P. A. and GRAHAM, M. E. (1951*a*). Remarks on low-aspect-ratio configurations in supersonic flow. *J. Aero. Sci.* **18**, 91–6.

LAGERSTROM, P. A. and GRAHAM, M. E. (1951*b*). Methods for calculating the flow in the Trefftz-plane behind supersonic wings. *J. Aero. Sci.* **18**, 179–90.

LAGERSTROM, P. A. and VAN DYKE, M. D. (1949). General considerations about planar and non-planar lifting systems. *Rep. Douglas Aircraft Co.* no. SM-13432.

LAITONE, E. V. (1947*a*). Subsonic flow about a body of revolution. *Quart. Appl. Math.* **5**, 227–31.

LAITONE, E. V. (1947*b*). The linearized subsonic and supersonic flow about inclined slender bodies of revolution. *J. Aero. Sci.* **14**, 631–42.

LANCE, G. N. (1952). The drag on slender pointed bodies in supersonic flow. *Quart. J. Mech. Appl. Math.* **5**, 165–77.

LAWRENCE, H. R. (1951). The lift distribution on low aspect ratio wings at subsonic speeds. *J. Aero. Sci.* **18**, 683–95.

LAWRENCE, T. (1952). Charts of the wave drag of wings at zero lift. *Curr. Paper Aero. Res. Coun., Lond.*, no. 116.

LESLIE, D. C. M. (1952). Supersonic theory of downwash fields. *Quart. J. Mech. Appl. Math.* **5**, 292–300.

LIGHTHILL, M. J. (1944*a*). The supersonic theory of wings of finite span. *Rep. Memor. Aero. Res. Coun., Lond.*, no. 2001.

LIGHTHILL, M. J. (1944*b*). A note on supersonic biplanes. *Rep. Memor. Aero. Res. Coun., Lond.*, no. 2002.

LIGHTHILL, M. J. (1945). Supersonic flow past bodies of revolution. *Rep. Memor. Aero. Res. Coun., Lond.*, no. 2003.

LIGHTHILL, M. J. (1948*a*). Supersonic flow past slender pointed bodies of revolution at yaw. *Quart. J. Mech. Appl. Math.* **1**, 76–89.

LIGHTHILL, M. J. (1948*b*). Supersonic flow past slender bodies of revolution, the slope of whose meridian section is discontinuous. *Quart. J. Mech. Appl. Math.* **1**, 90–102.

LIGHTHILL, M. J. (1949*a*). The flow behind a stationary shock. *Phil. Mag.* (7), **40**, 214–20.

LIGHTHILL, M. J. (1949*b*). A technique for rendering approximate solutions to physical problems uniformly valid. *Phil. Mag.* (7), **40**, 1179–1201.

LIGHTHILL, M. J. (1949*c*). The drag integral in the linearized theory of compressible flow. *Quart. J. Math.* (Oxford), **20**, 121–3.

LIGHTHILL, M. J. (1951). A new approach to thin aerofoil theory. *Aero. Quart.* **3**, 193–210.

LOMAX, H. and BYRD, P. F. (1951). Theoretical aerodynamic characteristics of a family of slender wing-tail-body combinations. *Tech. Notes Nat. Adv. Comm. Aero., Wash.*, no. 2554.

LOMAX, H. and HEASLET, M. A. (1949). Linearized lifting-surface theory for swept-back wings with slender planforms. *Tech. Notes Nat. Adv. Comm. Aero., Wash.*, no. 1992.

LOMAX, H. and HEASLET, M. A. (1951). Generalized conical-flow fields in supersonic theory. *Tech. Notes Nat. Adv. Comm. Aero., Wash.*, no. 2497.

LOMAX, H., HEASLET, M. A. and FULLER, F. B. (1951). Integrals and integral equations in linearized wing theory. *N.A.C.A. Rep.* no. 1054.

LOMAX, H. and SLUDER, L. (1952). Chordwise and compressibility corrections to slender-wing theory. *N.A.C.A. Rep.* no. 1105.

LOMAX, H., SLUDER, L. and HEASLET, M. A. (1950). The calculation of downwash behind supersonic wings with an application to triangular planforms. *N.A.C.A. Rep.* no. 957.

LUDLOFF, H. F. and REICHE, F. (1949). Linearized treatment of supersonic flow through ducts. *J. Aero. Sci.* **16**, 5–21.

MALVESTUTO, F. S. and MARGOLIS, K. (1950). Theoretical stability derivatives of thin sweptback wings tapered to a point with sweptback or sweptforward trailing edges for a limited range of supersonic speeds. *N.A.C.A. Rep.* no. 971.

MALVESTUTO, F. S., MARGOLIS, K. and RIBNER, H. S. (1950). Theoretical lift and damping in roll at supersonic speeds of thin sweptback tapered wings with streamwise tips, subsonic leading edges, and supersonic trailing edges. *N.A.C.A. Rep.* no. 970.

MARTIN, J. C. (1950). The calculation of downwash behind wings of arbitrary planform at supersonic speeds. *Tech. Notes Nat. Adv. Comm. Aero., Wash.*, no. 2135.

MARTIN, J. C. (1952). A vector study of linearized supersonic flow. Applications to non-planar problems. *Tech. Notes Nat. Adv. Comm. Aero., Wash.*, no. 2641.

MASLEN, S. H. (1952). Supersonic conical flow. *Tech. Notes Nat. Adv. Comm. Aero., Wash.*, no. 2651.

MEYER, R. E. (1948). The method of characteristics for problems of supersonic flow involving two independent variables—Part II: Integration along characteristics, the radial focusing effect in axially symmetrical flow. *Quart. J. Mech. Appl. Math.* **1**, 451–69.

MILES, J. W. (1950). A note on supersonic flow in the Trefftz plane. *Quart. Appl. Math.* **7**, 470–2.

MILES, J. W. (1952). On interference factors for finned bodies. *J. Aero. Sci.* **19**, 287.

MILES, J. W. (1953). Virtual momentum and slender body theory. *Quart. J. Mech. Appl. Math.* **6**, 286–9.

MIRELS, H. (1948). Theoretical wave drag and lift of thin supersonic ring airfoils. *Tech. Notes Nat. Adv. Comm. Aero., Wash.*, no. 1678.

MIRELS, H. (1951). A lift-cancellation technique in linearized supersonic-wing theory. *N.A.C.A. Rep.* no. 1004.

MIRELS, H. and HAEFELI, R. C. (1950). The calculation of supersonic downwash using line vortex theory. *J. Aero. Sci.* **17**, 13–21.

MOORE, F. (1950). Linearized supersonic axially symmetrical flow about open-nosed bodies obtained by use of stream function. *Tech. Notes Nat. Adv. Comm. Aero., Wash.*, no. 2116.

MORIKAWA, G. K. (1949). The wing-body problem for linearized supersonic flow. Thesis, Cal. Inst. Tech.

MORIKAWA, G. K. (1951). Supersonic wing-body lift. *J. Aero. Sci.* **18**, 217–28.

MORIKAWA, G. K. (1952*a*). A non-planar boundary problem for the wave equation. *Quart. Appl. Math.* **10**, 129–40.

MORIKAWA, G. K. (1952*b*). Supersonic wing-body-tail interference. *J. Aero. Sci.* **19**, 333–40.

MOSKOWITZ, B. (1951). Approximate theory for calculation of lift of bodies, afterbodies, and combinations of bodies. *Tech. Notes Nat. Adv. Comm. Aero., Wash.*, no. 2669.

MULTHOPP, H. (1951). A unified theory of supersonic wing flow, employing conical fields. *R.A.E. Rep. Aero.*, no. 2415.

MUNK, M. M. (1924). The aerodynamic forces on airship hulls. *N.A.C.A. Rep.* no. 184.

MUNK, M. M. (1950). The reversal theorem of linearized supersonic airfoil theory. *J. Appl. Phys.* **21**, 159–61.

NEUMARK, S. (1950). Critical Mach numbers for swept-back wings. *Aero. Quart.* **2**, 85–110.

NIELSON, J. N. and PITTS, W. C. (1952). Wing-body interference at supersonic speeds with an application to combinations with rectangular wings. *Tech. Notes Nat. Adv. Comm. Aero., Wash.*, no. 2677.

NONWEILER, T. R. F. (1950). Theoretical supersonic drag of non-lifting infinite span wings. *Rep. Memor. Aero. Res. Coun., Lond.*, no. 2795.

NONWEILER, T. R. F. (1951*a*). Theoretical stability derivatives of a highly swept delta wing and slender body configuration. *Rep. Coll. Aero., Cranfield*, no. 50.

NONWEILER, T. R. F. (1951*b*). Theoretical lift and pitching moment of a highly-swept delta wing on a body of elliptic cross-section. *Curr. Paper Aero. Res. Coun., Lond.*, no. 58.

NONWEILER, T. R. F. (1953). Theoretical wave drag at zero lift of fully tapered swept wings of arbitrary section. *Rep. Coll. Aero., Cranfield*, no. 76.

O'KEEFFE, J. (1952). The direct use of Green's method for supersonic potentials. *Quart. J. Mech. Appl. Math.* **5**, 82–92.

PRANDTL, L. (1904). Über die stationären Wellen in einem Gasstrahl. *Phys. Z.* **5**, 599–602.

PRANDTL, L. (1930). Über Strömungen deren Geschwindigkeiten mit der Schallgeschwindigkeit vergleichbar sind. *J. Aero. Res. Inst. Tokyo*, **63**, 14 et seq.

PUCKETT, A. E. (1946). Supersonic wave drag of thin airfoils. *J. Aero. Sci.* **13**, 475–84.

PUCKETT, A. E. and STEWART, H. J. (1947). Aerodynamic performance of delta wings at supersonic speeds. *J. Aero. Sci.* **14**, 567–78.

REISSNER, E. (1949). On compressibility corrections for subsonic flow over bodies of revolution. *Tech. Notes Nat. Adv. Comm. Aero., Wash.*, no. 1815.

RIBNER, H. S. (1947). The ring airfoil in non-axial flow. *J. Aero. Sci.* **14**, 529–30.

RIBNER, H. S. (1950*a*). On the effect of subsonic trailing edges on damping in roll and pitch of thin sweptback wings in a supersonic stream. *Tech. Notes Nat. Adv. Comm. Aero., Wash.*, no. 2146.

RIBNER, H. S. (1950*b*). Some conical and quasi-conical flows in linearized supersonic wing theory. *Tech. Notes. Nat. Adv. Comm. Aero., Wash.*, no. 2147.

RIBNER, H. S. and MALVESTUTO, F. S. (1948). Stability derivatives of triangular wings at supersonic speeds. *N.A.C.A. Rep.* no. 908.

ROBINSON, A. (1946*a*). Wave drag of diamond shaped aerofoils at zero incidence. *Rep. Memor. Aero. Res. Coun., Lond.*, no. 2394.

ROBINSON, A. (1946*b*). Aerofoil theory of a flat delta wing at supersonic speed. *Rep. Memor. Aero. Res. Coun., Lond.*, no. 2548.

ROBINSON, A. (1948*a*). Rotary derivatives of a delta wing at supersonic speeds. *J. Roy. Aero. Soc.* **52**, 735–52.

ROBINSON, A. (1948*b*). On source and vortex distributions in the linearized theory of steady supersonic flow. *Quart. J. Mech. Appl. Math.* **1**, 408–32.

ROBINSON, A. (1952). Aerofoil theory for swallow-tail wings of small aspect ratio. *Aero. Quart.* **4**, 69–82.

ROBINSON, A. and HUNTER-TOD, J. H. (1947*a*). Bound and trailing vortices in the linearized theory of supersonic flow, and the downwash in the wake of a delta wing. *Rep. Coll. Aero., Cranfield*, no. 10.

ROBINSON, A. and HUNTER-TOD, J. H. (1947*b*). Aerodynamic derivatives with respect to sideslip for a delta wing with small dihedral at supersonic speeds. *Rep. Coll. Aero., Cranfield*, no. 12.

RODRIGUEZ, A. M., LAGERSTROM, P. A. and GRAHAM, E. W. (1954). Theorems concerning the drag reduction of wings of fixed planform. *J. Aero. Sci.* **21**, 1–7.

ROPER, G. M. (1948). The flat delta wing at incidence, at supersonic speeds, when the leading edges lie outside the Mach cone of the vertex. *Quart. J. Mech. Appl. Math.* **1**, 327–43.

ROPER, G. M. (1949*a*). The yawed delta wing at incidence at supersonic speeds. *Quart. J. Mech. Appl. Math.* **2**, 354–73.

ROPER, G. M. (1949*b*). The pressure distribution, at supersonic speeds and zero lift, on some sweptback wings having symmetrical sections with rounded leading edges. *Rep. Memor. Aero. Res. Coun., Lond.*, no. 2700.

SACKS, A. H. (1952). Behaviour of vortex systems behind cruciform wings—motions of fully rolled up vortices. *Tech. Notes Nat. Adv. Comm. Aero., Wash.*, no. 2605.

SCHLICHTING, H. (1936). Tragflugeltheorie bei Überschallgeschwindigkeit. *Luftfahrtforschung*, **13**, 320–35. (Transl. *Tech. Memor. Nat. Adv. Comm. Aero., Wash.*, no. 897.)

SEARS, W. R. (1946). On compressible flow about bodies of revolution. *Quart. Appl. Math.* **4**, 191–3.

SEARS, W. R. (1947). A second note on compressible flow about bodies of revolution. *Quart. Appl. Math.* **5**, 89–91.

SEARS, W. R. (1949). The linear-perturbation theory for rotational flow. *J. Math. Phys.* **28**, 268–71.

SEARS, W. R. and TAN, H. S. (1951). The aerodynamics of supersonic biplanes. *Quart. Appl. Math.* **9**, 67–76.

SHAW, B. W. B. (1950). Nose controls on delta wings at supersonic speeds. *Rep. Coll. Aero., Cranfield*, no. 36.

SIBERT, H. W. (1947). Approximations involved in the linear differential equation for compressible flow. *J. Aero. Sci.* **14**, 680–1.

SNOW, R. M. (1948). Aerodynamics of thin quadrilateral wings at supersonic speeds. *Quart. Appl. Math.* **5**, 417–28.

SPREITER, J. R. (1950). The aerodynamic forces on slender plane- and cruciform-wing and body combinations. *N.A.C.A. Rep.* no. 962.

SPREITER, J. R. and SACKS, A. H. (1951). The rolling up of the trailing vortex sheet and its effect on the downwash behind wings. *J. Aero. Sci.* **18**, 21–32.

SQUIRE, H. B. (1947). An example in wing theory at supersonic speeds. *Rep. Memor. Aero. Res. Coun., Lond.*, no. 2549.

STEWART, H. J. (1946). The lift of a delta wing at supersonic speeds. *Quart. Appl. Math.* **4**, 246–54.

STEWARTSON, K. (1950). Supersonic flow over an inclined wing of zero aspect ratio. *Proc. Camb. Phil. Soc.* **46**, 307–15.

STOCKER, P. M. (1951). Supersonic flow past bodies of revolution with thin wings of small aspect ratio. *Aero. Quart.* **3**, 61–79.

TAUNT, D. R. and WARD, G. N. (1946*a*). The axially symmetrical supersonic free jet. (Unpublished Admiralty report.)

TAUNT, D. R. and WARD, G. N. (1946*b*). Wings of finite aspect ratio at supersonic velocities. *Rep. Memor. Aero. Res. Coun., Lond.*, no. 2421.

TAYLOR, G. I. (1932). Applications to aeronautics of Ackeret's theory of aerofoils moving at speeds greater than that of sound. *Rep. Memor. Aero. Res. Coun., Lond.*, no. 1467.

TSIEN, H. S. (1938). Supersonic flow over an inclined body of revolution. *J. Aero. Sci.* 5, 480–3.

TSIEN, H. S. and LEES, L. (1945). The Glauert-Prandtl approximation for subsonic flows of a compressible fluid. *J. Aero. Sci.* 12, 173–87.

TUCKER, W. A. and NELSON, R. L. (1949). Theoretical characteristics in supersonic flow of two types of control surfaces on triangular wings. *N.A.C.A. Rep.* no. 939.

URSELL, F. and WARD, G. N. (1950). On some general theorems in the linearized theory of compressible flow. *Quart. J. Mech. Appl. Math.* 3, 326–48.

VAN DYKE, M. D. (1951). First- and second-order theory of supersonic flow past bodies of revolution. *J. Aero. Sci.* 18, 161–78.

WALCHNER, O. (1937). Zur Frage der Widerstandsverringerung von Tragflügeln bei Überschallgeschwindigkeit durch Doppeldeckeranordnung. *Luftfahrtforschung*, 14, 55–62.

WARD, G. N. (1945). A note on compressible flow in a tube of slightly varying cross-section. *Rep. Memor. Aero. Res. Coun., Lond.*, no. 2183.

WARD, G. N. (1946). The pressure distribution on some flat laminar aerofoils at incidence at supersonic speeds. *Rep. Memor. Aero. Res. Coun., Lond.*, no. 2206.

WARD, G. N. (1948). The approximate external and internal flow past a quasi-cylindrical tube moving at supersonic speeds. *Quart. J. Mech. Appl. Math.* 1, 225–45.

WARD, G. N. (1949*a*). Supersonic flow past slender pointed bodies. *Quart. J. Mech. Appl. Math.* 2, 75–97.

WARD, G. N. (1949*b*). Calculation of downwash behind a supersonic wing. *Aero. Quart.* 1, 35–8.

WARD, G. N. (1949*c*). Supersonic flow past thin wings. I. General theory. *Quart. J. Mech. Appl. Math.* 2, 136–52.

WARD, G. N. (1949*d*). Supersonic flow past thin wings. II. Flow reversal theorems. *Quart. J. Mech. Appl. Math.* 2, 374–84.

WARD, G. N. (1950). The supersonic flow past a slender ducted body of revolution with an annular intake. *Aero. Quart.* 1, 305–18.

WARD, G. N. (1952*a*). On the integration of some vector differential equations. I. *Quart. J. Mech. Appl. Math.* 5, 432–40.

WARD, G. N. (1952*b*). On the integration of some vector differential equations. II. Application to the linearized theory of steady compressible fluid flow. *Quart. J. Mech. Appl. Math.* 5, 441–6.

WHITHAM, G. B. (1950). The behaviour of supersonic flow past a body of revolution, far from the axis. *Proc. Roy. Soc.* A, 201, 89–109.

WHITHAM, G. B. (1952). The flow pattern of a supersonic projectile. *Commun. Pure Appl. Math.* 5, 301–48.

YOUNG, A. D. and KIRKBY, S. (1947). Applications of the linear perturbation theory of compressible flow about bodies of revolution. *Rep. Memor. Aero. Res. Coun., Lond.*, no. 2624.

INDEX